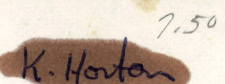

BIOLOGY OF SEX

BIOLOGY OF SEX

Charlotte J. Avers
Rutgers University

John Wiley & Sons, Inc., New York · London · Sydney · Toronto

To Ron

Copyright © 1974, by John Wiley & Sons, Inc.

All rights reserved. Published simultaneously in Canada.

No part of this book may be reproduced by any means, nor transmitted, nor translated into a machine language without the written permission of the publisher.

Library of Congress Cataloging in Publication Data:

Avers, Charlotte J
   Biology of sex.

   Includes bibliographies.
   1. Sex (Biology)   2. Human reproduction.
3. Human genetics.   I. Title. [DNLM: 1. Reproduction.
2. Sex. QH481 A953b 1974]

QP251.A9   612.6   74-1021
ISBN 0-471-03842-3

Printed in the United States of America

10 9 8 7 6 5 4 3 2 1

Preface

A variety of shorter and more topically restricted textbooks has appeared in response to the changing curricula of colleges and universities in recent years. In part, this reflects the diverse organization of introductory courses and in part the availability of introductory level courses that emphasize selected aspects of a subject rather than the more inclusive surveys. It should be possible for instructors to create more individually tailored courses and use two or more shorter texts covering different topics within a discipline, or to offer short courses in which one topic is emphasized and others either excluded or discussed peripherally to the major theme. This textbook is intended for introductory level courses in which human biology is the focus for discussions of general biological principles and thinking, particularly for the theme of sex as a biological phenomenon. It could also be used as a text along with others for courses that survey a set of themes in a more comprehensive study of Biology.

It is possible only to consider the subject of sex in humans but, in my opinion, this approach ignores the universality of the phenomenon in living systems. We are part of a biological continuum, and we share an evolutionary history with other forms of life on this planet. Not only are there threads of continuity that bind together the life of past and present times but, very often, we know more about sexual systems in other species and can profit from the information and comparisons with ourselves. I believe that we will gain a deeper understanding of human traits if these are seen in perspective with the differences and similarities displayed by other kinds of organisms.

This book is organized in a fashion that reflects my own prejudices in a sequential presentation, but it should be possible to discuss some chapters in alternative sequences from those that are given here. If a genetics background is unnecessary for comprehending human reproduction, then there is no reason why Part II could not be studied first, and Part I at a later time. The chapters in Part I are intended to introduce the student to the general premises of genetics, the particular aspects of inheritance that bear more directly on an understanding of the human system, and to comparisons and summarizations of these topics for a variety of systems, in the context of their evolutionary development through time.

Part II is concerned specifically with human reproduction. The basic anatomy and physiology of males and females forms the basis for understanding the events and phenomena that contribute to the formation and development of new members of the species in each generation. Once these matters have been described, it becomes possible to pursue the subject of fertility in humans and to direct the discussion toward the methods of control that we can exert, alone among all the species on the Earth.

Part III covers topics that most directly influence the activities of human beings as social creatures. General themes and kinds of behavior are studied as a prerequisite to understanding human behavior in the context of an animal species, but as a very special kind of animal who can manipulate his or her existence by virtue of a high level of intelligence. There can be a greater appreciation of human behavior if its similarities and its uniqueness are viewed relative to other mammalian species. The element of control that we can exert in managing our existence is further emphasized in the final two chapters. Do we play out our societal and interpersonal roles as a result of proscribed genetic instructions or have we the ability to modulate our behavior with the tools of reasoning and thinking? Can we manage to regulate our population size as well as other elements of our biological potentials? I would like to believe that you will evaluate the problems discussed in the final chapters in a more comprehensive and nonstereotyped way because of a greater understanding and appreciation of the human individual gained from the topical treatments presented in the earlier parts of the book.

A brief list of readings has been provided for each chapter for those who wish to read further about some topic, or to evaluate a subject independently after reading the opinions of other authors. Each of the brief chapters only touches upon a subject and, for some students, this may not be satisfying. Books and instructors teach; it is the student who learns as much or as little as he or she may wish to extract from such teaching.

Charlotte J. Avers

New Brunswick, N.J., 1973

Contents

part one
GENETICS AND EVOLUTION   1

chapter one
Basic Genetic Mechanisms   3
    Introduction   3
    The Hereditary Machinery   6
       Chromosomes   7
       The Nature and Variety of Genes   10
    Mitosis and Meiosis   18
       Mitosis   20
       Meiosis   21
    The Determination and Differentiation of Sex   25
    Suggested Readings   28

chapter two
Sex-Related Inheritance Patterns   29
    Patterns of Inheritance   29
       Autosomal Inheritance   30

CONTENTS

      Sex-Linked Inheritance   31
      Other Inheritance Patterns Related to Sex   35
    Pedigree Analysis   38
      Autosomal Patterns   39
      Sex-Linked Inheritance   42
    Suggested Readings   47

chapter three

Human Medical Genetics   48

    The Human Chromosomes   48
      The Sex Chromosomes   49
      Autosomal Chromosome Anomalies   56
      Significance of Chromosomal Anomalies   62
    Medical Genetics   63
      Determination of Risk   64
      Prenatal Diagnosis   66
    Genetic Counseling   68
    Suggested Readings   69

chapter four

Evolutionary Considerations   71

    Some Evolutionary Principles   71
      Origins of Diversity   72
      Natural Selection and Adaptation   77
    The Evolution of Sex   78
      Evolution of Gametes   78
      Haploid to Diploid   81
      Protection of the Egg and the Embryo   82
    Sex Determination   85
      Systems of Determination   85
      Intactness of the Sex Chromosomes   87
    Suggested Readings   87

part two

HUMAN REPRODUCTION   89

chapter five

Reproductive Anatomy and Physiology   91

    General Features   91
    The Male Reproductive System   93
      Reproductive Anatomy   94

Hormonal Control	101
The Female Reproductive System	106
Reproductive Anatomy	107
Growth of the Ovum and Follicle	108
Formation of the Corpus Luteum	110
Uterine Changes and Menstruation	114
General Effects of Female Sex Hormones	115
The Estrous Cycle and Fertility	115
Development of Biological Sex	117
Suggested Readings	121

chapter six
Fertilization, Pregnancy, Birth, and Lactation   122

Fertilization and Implantation	123
Fusion of the Egg and Sperm	125
Implantation	126
The Placenta	127
Hormonal Changes During Pregnancy	127
Development of the Embryo and Fetus	129
The First Eight Weeks: The Embryo	131
The Third Through the Sixth Months	135
The Final Trimester	137
Difficulties During Pregnancy	138
Birth	141
Labor	141
Multiple Births	142
Hormonal Physiology	143
Lactation	144
Anatomy and Physiology	144
Lactose and Lactase	147
Suggested Readings	148

chapter seven
Fertility and Infertility in Humans   149

Methods of Fertility Control	150
Sexual Abstinence	150
Contraception	152
Sterilization	159
Abortion	160
Infertility	162
Causes of Childlessness	162
Remedies for Childlessness	163
Suggested Readings	165

part three
SOCIAL AND REPRODUCTIVE BEHAVIORS — 168

chapter eight
General Aspects of Animal Behavior — 169

- Features of the System — 169
- Stereotyped Behaviors — 171
    - Reflex Behavior — 172
    - Spontaneous Activity — 173
    - Fixed-Action Patterns — 174
    - Taxis and Tropism — 175
- Motivated Behaviors — 175
    - Neural Influences — 176
    - Pattern Periodicities — 179
    - Hormonal Influences — 180
    - Experiential Influences — 181
    - Practice and Experience — 182
    - Imitation — 183
- Learning — 183
    - Associative Learning — 184
    - Reasoning — 185
- Suggested Readings — 186

chapter nine
Sexual and Sex-Related Social Behavior — 187

- Social Order — 187
    - Hierarchies of Dominance — 188
    - Territoriality — 189
- Social Organization — 190
    - Insect Societies — 190
    - Types of Societies in Mammals — 192
- Mating Behavior — 194
    - Meeting of the Sexes — 194
    - Coordination of Reproductive States — 195
    - Care of the Young — 196
- Development of Behavior — 197
    - Interactions Between Mother and Infant — 198
    - Interactions Among Peers — 200
    - Paternal Love — 200
- Suggested Readings — 202

chapter ten
Human Sexual Behavior — 203
- Measurements Versus Assessments — 204
  - Statistical Information — 205
  - Physiological Concomitants — 210
- Varieties of Human Sexual Experience — 213
  - Comparative Studies — 214
- Variations in Human Sexual Behavior — 217
  - Variations in Object Choice — 217
  - Variations in Sexual Aim — 220
- Diseases Involving Reproductive Organs — 221
  - Venereal Diseases — 221
    - Cancers of Reproductive Organ Systems — 223
- Suggested Readings — 224

chapter eleven
Sex Roles in Society — 226
- Human Societies — 227
  - Pattern Variability — 227
  - Care of the Young — 228
- The Big Hunting Mutation — 229
  - Principal Lines of Evidence — 232
  - Cross-Cultural Human Patterns — 244
  - Behavioral Plasticity — 245
- Development of Gender Dimorphism — 246
  - Development of Biological Sex — 246
  - Clinical Studies — 247
- Suggested Readings — 249

chapter twelve
Population Dynamics — 251
- Birth Rates — 251
- Death Rates — 254
- Population Growth — 255
  - Factors Affecting Population Growth — 259
  - Some Current Concerns About Overpopulation — 260
- Some Prospects for the Future — 264
  - Prospects for Improvement — 265
  - For How Long Will We Last? — 267
- Suggested Readings — 268

Index — 271

part one
GENETICS AND EVOLUTION

chapter one

Basic Genetic Mechanisms

Introduction

As an introduction to an examination of sex as a biological phenomenon, we might ask these general questions: What is sex? When did sex appear during the evolution of life forms on the Earth? Which kinds of life exhibit sexual activities? And, finally, why sex? These questions immediately imply that alternatives to sex do exist among present-day species and that sex was not a feature of the first forms of life that populated our planet during its earliest history.

What is sex? In its basic context, sex is a reproductive mechanism in which two parent cells fuse to produce a new individual of the next generation. In its more familiar form, sexual reproduction involves the fusion of an egg cell, produced by the female parent, with the sperm cell produced by the male parent. The product of cell fusion is called a fertilized egg because the particular fusion has been designated as *fertilization.* Actually, many variations on this basic theme have been described. The sex cells that are capable of fusion, generically called *gametes*, may be similar in size, shape, and behavior or be as different from each other as a sperm and an egg. The gametes may be produced by a single individual who possesses organs for sperm and for egg production or by separate individuals in the classical *biparental reproduction.* Numerous other variables also have been found among the hundreds of thousands of sexually reproducing species that are known.

When did sex arise during evolution? Although we don't have the formal answer to this, we can make educated guesses on the basis of comparative information derived from fossil evidence and from descendant species that are presently living on the Earth. During the first two billion years after life originated, simple bacterialike forms and other types were the only sort that were present. The living relatives of these ancient species are lacking in conventional sexual reproductive capacity; that is, reproduction is primarily *asexual*. The logical deduction is that sex did not exist during this interval in evolution and that the continuity of life resided in species reproduction by asexual means. About one billion or more years ago, a new life form evolved. The new form contained a very different organization of cellular structure than had existed, and still exists, in the progenitor species. One of the fundamental changes involved the organization of membrane-bound internal structures, especially of the nucleus and its gene content. We recognize these two great divisions of life forms today as *prokaryotes,* lacking a true membrane-bound nucleus, and *eukaryotes,* which do possess a conventional nucleus, chromosomes, and gene arrangements on these chromosomes. Some time after the appearance of eukaryotes, various evolutionary changes were incorporated that led to the development of sexual reproductive capacities. The great explosion of new life forms that characterizes the last billion years of evolution is believed to be due considerably to the introduction of sexual reproduction in eukaryotes.

Which kinds of life exhibit sexual activities? As just mentioned, eukaryotes are sexual and prokaryotes are not. However, there are quite a few eukaryote species which have lost sexual capacity or perhaps never have evolved such a mechanism during their long history on the Earth. Some of the simple life forms such as ameba are asexual, but close relatives of ameba do possess sexual reproduction ability. Many familiar species of flowering plants reproduce asexually, including the dandelions that pepper our lawns despite the brave efforts of the herbicide manufacturers. In most cases, we are fairly sure that asexuality has replaced sexual reproduction by genetic and evolutionary events. In other cases, we don't know whether sex has been lost or if perhaps it never existed in a particular group of organisms. All animals with backbones (vertebrates) are exclusively sexual in reproduction, including the fishes and other aquatic groups, amphibians, reptiles, birds, and the mammals to which our own species belongs. All of the vertebrates also show biparental reproduction, the sperm being produced by males and the eggs by females. Invertebrate animals exhibit many variations on this theme, as do microorganisms and the several groups of plants.

Why sex? At first glance, it is puzzling to explain the enormous success of sexual species because asexual reproduction is far more economical and prolific. A single bacterial cell may produce billions of offspring in a 24-hour day by the simple process of the form of cell division known as fission. If we place one or a few bacteria into a clear broth and examine the liquid, the next day we can see the cloudiness which indicates the teeming bacterial population that is inside the container. An orange seems to become moldy within a day or two, these greenish growths representing the billions of new mold cells that have been produced by one or a few initial mold cells. Sexually reproducing species sometimes are prolific too, but usually they produce fewer offspring or only a few survive the first hours or days of their existence. Millions of sperm are required for an egg to be fertilized successfully, all of the sperm but one being effectively wasted. Considering the lower efficiency of sexual reproduction as compared with asexual mechanisms, how can we explain the institution of sex during evolution and its obvious success, if we are to judge by the fossil and living record of life? The answer to these questions is genetic; sex permits the *recombination of genes* and thus of offspring which are slightly different from their parents and from each other. Except for brothers or sisters who were produced from the same fertilized egg, such as identical twins, no two human beings are exactly alike genetically. The variability that is obvious within our own populations and families is also typical of other sexually reproducing species, whether or not we can distinguish one chicken from another or perceive the differences in unfamiliar groups of organisms in general. Careful study quickly reveals the wealth of variation within a sexual species. Gene recombination leads to a fantastic level of inherited variation in populations, and variability is the raw material upon which evolutionary progress depends. Sexual reproduction has been remarkably successful and has become the sole reproductive process in most species by virtue of the variability that is possible and of the evolutionary advantages of such variability. The proliferation and diversity of life forms during the past billion years and the continuing improvement in more recent species groups can be traced directly to the capacity for gene recombination that is permitted by sex. The first four chapters will explore these topics in greater detail.

Sex is an inherited property of many kinds of plant and animal life, from simple one-celled forms to the most advanced and complex species. The genes that are responsible for our sexual characteristics exist in chromosomes in the cell nucleus, but they represent stored information that must be translated into the final chemical and structural materials that make up an individual. The set of genes represents a series of blueprints that contain the specifications and directions for the development of an organism, whether human or microbe. Genetic information is stored potential, and it can be mobilized by the chemical and structural machinery of cells to yield very specific products that distinguish humans from microbes and one human from another. But if we stop to think about this for a minute, we realize that each one of us is made up of many different kinds of cells and tissues. If we display variously formed and functioning cells, does this mean that there are different genes in different cells of a single individual? Are all of our characteristics developed exclusively according to genetic instructions or can these instructions be modulated so that slightly different products may be formed under the influence of environmental agencies? Can we modulate and influence the genes we contain to produce characteristics that are more desirable or beneficial to an individual?

These questions and many others have been answered in recent years as biologists began to probe even more deeply into fundamental properties of cell chemistry and function. During the next few chapters, we will consider these types of questions and lay the groundwork for a more effective understanding of what we are and how we compare with millions of other forms of life sharing this planet. If we are to understand ourselves and our unique qualities, it is essential first to explore the genetic system that directs the development of life forms. Once this foundation has been established, then we should be in a better position to evaluate the expressions of our genes in the development of our biological properties, including sex.

The Hereditary Machinery

The thread of continuity that links generations exists in genes of the individual and the species. We resemble our parents more than we resemble other people because we share more genes with them than with the rest of the population. But we all are recognizable as human beings and not as some other kind of animal because we share a common pool of genes that is different from the gene pools of other species. At the same time, we must have many genes in common with other organisms because there are similarities that bind us together in an evolutionary series of relatives, some more distant than others. As you wander through a zoo, it is very obvious that the monkeys and apes are more like each other than like lions and tigers, and that monkeys and apes resemble us more closely than any other form of life. Here again, we see the sharing of genes from countless past generations of evolutionary history. But there are differences among monkeys, apes, and humans, which we know to be due to changes in genes during evolution. Still, this doesn't tell us whether every characteristic of the animal is the direct expression of its genes. The environment in which the animal lives is an important element in the final expression of some of its genes. A chimpanzee that is well fed may be larger and more active than one that has been maltreated, even though the genetic instructions specified that each animal was to be of a certain size and temper. In many cases, it has been clearly established that the genetic potential for development may be influenced by the conditions under which this potential is expressed. There is no "nature versus nurture" controversy as many popular magazines continue to insist. Except for a few of our chemical characteristics, the development of the individual depends upon the interaction of genetic (nature) and environmental (nurture) factors.

The potential of each individual is the sum of the genes received from its parents. In species that reproduce asexually, there is only one parent who contributes genes to the progeny of the next generation, and these progeny essentially are identical to the parent. For sexually reproducing species such as our own, each individual receives a combination of genes from two parents. The sperm and egg each contain one set of complementary genes and, when they fuse to form the new individual of the next generation, there is a different combination of genes than was present in either parent. But the difference is relatively slight so that each child closely resembles its parents. In fact, except for identical twins or other sibling groups, there are no two individuals who are exactly alike genetically. Later, we will see why this is true.

The blood type of an infant depends on the particular genes that were transmitted by its mother and father through the egg and sperm, respectively, and on the activity of these genes in the synthesis of the blood proteins that represent the infant's blood type. The proteins themselves are not inherited any more than the eyes or hair or color of the skin. Each trait develops according to the chemical pathways specified by the genes, but each is influenced to some degree by the internal environment in which this chemistry proceeds. The information for the manufacture of specific proteins is incorporated into the molecular structure of the gene, and it is this informational potential that is transmitted to each generation during reproduction, whether sexual or asexual. The expression of this potential takes place over the years of growth and development, and in many cases right up to the moment of death.

Chromosomes

The **genes** are contained within **chromosomes,** which are rather complex structures (Fig. 1.1) made up of various kinds of chemical materials. Patterns of inheritance can best be understood if we consider genes as

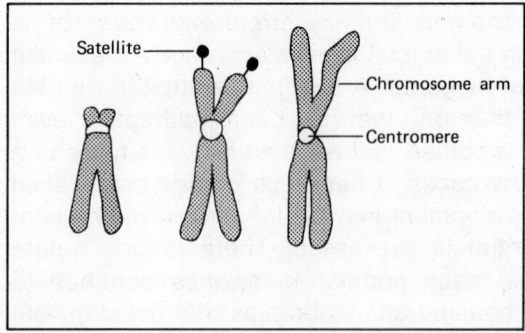

figure 1.1
Examples of chromosomes as they appear in dividing cells seen under the microscope. Each chromosome is a double structure; the two halves separate at a later stage when they become parts of separate daughter nuclei. Chromosome lengths vary, as do the arms of the chromosome on either side of the *centromere* (the region active in transport of chromosomes to opposite ends of the cell). In addition to these differences in size and form, some chromosomes can be further recognized by the presence of terminal knobs, called *satellites*.

integral parts of chromosomal structure. Thus we can follow transfer of genetic information from one generation to the next by analyzing the patterns of transmission of the chromosomes themselves. In most animals, we can distinguish two kinds of chromosomes, **autosomes** and **sex chromosomes,** by one or two criteria. Vertebrate animals have chromosomes that can be recognized according to microscopical features as being autosomes or sex chromosomes, principally on the basis of differences in their staining properties in cells that have been killed and sliced for observations with the conventional light microscope. Most of the simpler forms of life and the great majority of plants do not possess sex chromosomes. In addition to identifying the kinds of chromosomes according to their appearance under the microscope, it is possible to distinguish sex chromosomes and autosomes on the basis of different patterns of inheritance of the genes that are present in each type. In some cases, where it is difficult to see any differences using the microscope, it still is possible to determine whether or not sex chromosomes are present according to the criterion of the inheritance patterns of their genes.

Although gene transmission patterns are different for autosomes and sex chromosomes, this does not imply that all of the genes related to sex are restricted to the sex chromosomes. In fact, this is not the case at all. A variety of genes occurs on all chromosomes of the species, and it must

Basic Genetic Mechanisms

be obvious that sexually reproducing forms that lack sex chromosomes must have a scattering of all kinds of genes on autosomes, the only chromosome type that they possess. However, in humans and other species that have sex chromosomes there are genes directly related to the determination of sex that are grouped together on particular chromosomes. Such sex chromosomes are an integral part of the mechanism by which the sex of an individual is determined at the moment of fusion of the egg and sperm. The differentiation of sexual traits occurs later on during development of the embryo and maturation of the individual, but the pathways of development are laid out at the time the egg is fertilized. The realization of these pathways of developmental potential depends on the cooperative action of genes on all of the chromosomes during the lifetime of the individual, beginning with the moment of conception.

In mammals, the sex of the new individual, whether human or rabbit, is determined when the sperm fertilizes the egg (Fig. 1.2). If the fertilized egg contains two **X** chromosomes, then the individual eventually will differentiate as a female, and if there is one **X** and one **Y** chromosome,

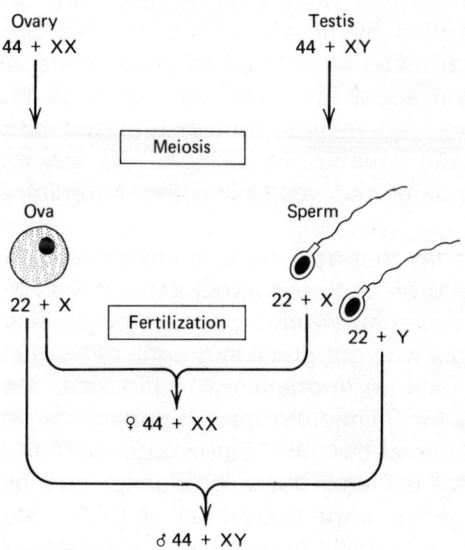

figure 1.2
The sex of the individual is determined at the time of fertilization in systems that have a sex chromosome mechanism. In humans and other mammals, the union of an egg with a sperm carrying an X chromosome leads to a female individual, whereas fertilization of an egg by a Y-carrying sperm determines that development will be male.

the development will lead to a male. The particular combination of the pair of sex chromosomes (**XX** or **XY**) determines sex, but the differentiation of sexual structures and functions is an outcome of the action of many genes over the span of time that is required for the embryo to achieve maturity as an adult. There are few absolutes, so that variations may develop depending on the conditions that prevail during the long period when the genetic potential is realized in the processes of differentiation. The differentiation of sex traits is influenced profoundly by the quality and quantity of sex hormones, but the potentials for hormone production and tissue development reside in the genetic information that is transmitted through the egg and sperm to the new generation.

The Nature and Variety of Genes

The fundamental rules of inheritance first were formulated by Gregor Mendel in 1865, but very little attention was paid to his work until 1900. By the turn of the century, a great deal had been learned about the cell that was unknown previously so that Mendel's important studies only could be appreciated after biologists had described chromosomes, fertilization of the egg by the sperm, and additional experiments on inheritance. Mendel's work was verified by others who also used the common garden pea as experimental material, as well as other kinds of organisms. Although little was known about the chemical nature of the gene until many years after 1900, the fundamental principles established by Mendel and subsequent geneticists continue to serve to this day as the solid foundation for understanding and analyzing the hereditary machinery of living systems.

Although we take it for granted today that genes are made up of DNA (deoxyribonucleic acid) molecules and we can read about the latest additions to our knowledge of DNA in newspapers and popular magazines, the first major line of evidence for this was not published until 1944. Just as Mendel's discoveries could not fully be appreciated in his time, the 1944 article was not directly useful for immediate new experiments, so this information also made no headlines at the time. But independent and complementary lines of experimental evidence were produced, and by 1952 it generally was agreed that genes were comprised of DNA. One reason for explaining the scientific community's resistance to accepting DNA as the genetic material is that it was believed that the molecule was essentially monotonous and incapable of accommodating the infinite variety that one would expect for genes. Since there are many different genes, their underlying chemistry had to be suitable to explain the great variety of observed genes. Many biologists and biochemists favored the idea that genes were protein molecules because proteins are known to be exceedingly diverse. In 1953, James D. Watson and Francis H. C. Crick published two brief articles that opened a new era of genetic analysis and contributed substantially to our basic understanding of the gene and its properties. Watson and Crick proposed a molecular model for DNA, and they discussed some of the properties of the gene that could be explained on the basis of this model. They suggested probable mechanisms by which the DNA molecule could replicate precisely, thus explaining the continuity between the generations; how mutations might occur, thus explaining the origin of new genes; and other important and thought-provoking features. Their proposals were so rich in experimental possibilities that they permitted a fantastic burst of studies, some of which are still in active progress more than twenty years later. Unlike the somewhat premature publications such as those of 1865 and 1944, the scientific community was geared up and ready to go once they read the 1953 Watson and Crick articles. The following brief discussions will summarize highlights of our genetic knowledge and how the hereditary system influences particular aspects of sex and other properties of living systems.

DNA, RNA, and Protein

The principal material of the gene is DNA in every biological system, but there are some viruses that have a genetic system that is based instead on RNA. The DNA molecule is remarkably versatile despite its simplicity of organization. Each molecule of DNA contains hundreds or thousands of nitrogen-containing bases, each of which is attached to a sugar group that forms part of a continuous sugar—phosphate "backbone." Two such linear strands generally are linked together by their bases in a particular way that forms the well known *double helix* three-dimensional structure (Fig. 1.3).

There are four kinds of bases in DNA: thymine (T), adenine (A), guanine (G), and cytosine (C), but there is only one kind of sugar and one kind of phosphate group that make up the twisting length of the molecule. When two of these strands intertwine to form a double helix, there is a restriction such that only T and A can bond together, and only G and C can bond together across the width of the molecule. Watson and Crick not only described this arrangement in their model, but they also suggested that the precision of DNA replication was due to this pairing specificity. When new DNA molecules are synthesized, the paired bases separate and each one can only be bonded to its special kind of partner when a new double helix is formed (Fig. 1.4). Thus, each new set of genes will be like the previous set because the replication of DNA molecules follows this particular pattern of manufacture. This mechanism is so successful that almost every form of life contains the same genetic chemicals and

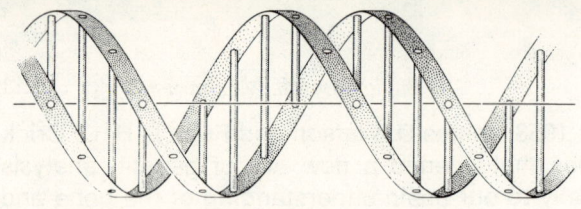

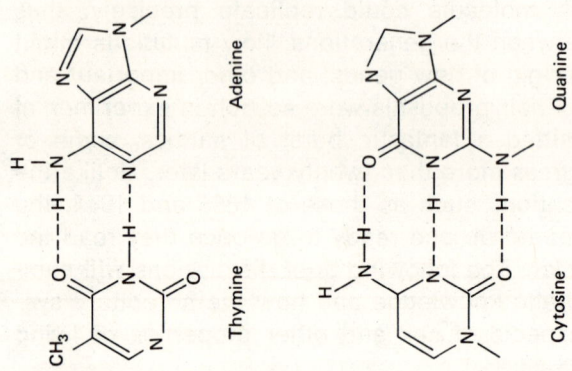

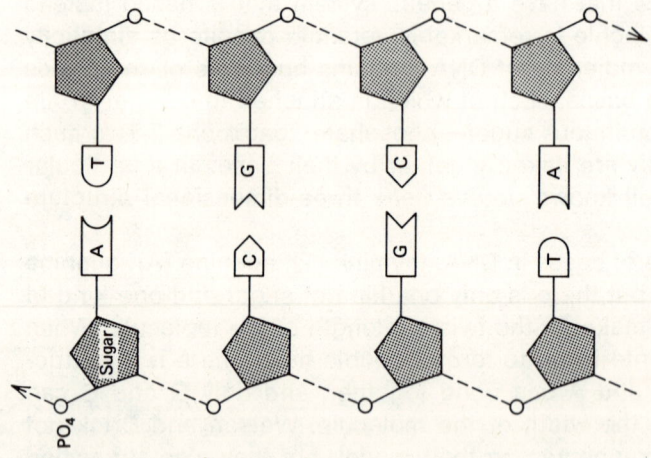

figure 1.3
The DNA molecules are duplex, each strand including a "backbone" of sugar and phosphate units in a vertical sequence, with lateral pairs of the bases Thymine, Adenine, Cytosine, and Guanine. (A) Thymine pairs only with adenine, and cytosine pairs only with guanine. (B) Any base pair may be stacked *vertically* adjacent to any other base pair throughout the length of the double helix. This contributes to DNA variety and gene diversity.

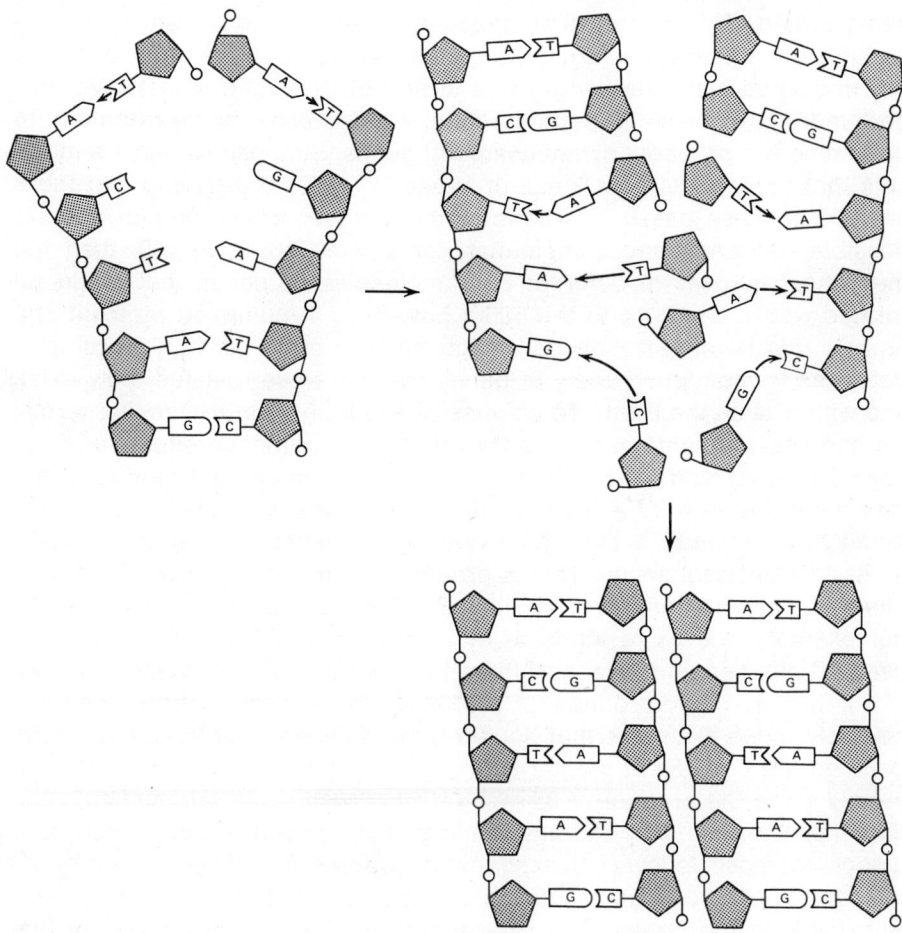

figure 1.4
The duplex DNA molecules replicate in a fashion that leads each original strand to direct the synthesis of a new strand that is *complementary* to itself. When this process is completed, there are two duplex molecules that are identical to each other and exactly like the originals from which they were fashioned. This mechanism helps to explain genetic continuity from generation to generation.

undergoes the same mode of replication of new genes. If DNA was the original genetic material at the dawn of life on this planet, then the same simple but effective machinery has been handed down through all the generations and changes during three billion years of evolutionary time.

Unlike the restriction in base pairing between different DNA strands, there is no known restriction on the vertical arrangement of bases in the *same* strand. This observation provides us with an important clue concerning the theoretical variety of genes. Since the bases can occur in any sequence along the length of a strand of DNA, and since there may be hundreds of bases in a single gene, we can use simple arithmetic to determine the probable diversity among genes. Suppose we have a molecule that contains all four kinds of bases (T, A, G, and C) and that there are 500 of these bases in a gene. If these bases are arranged in every possible sequence and combination for a total length of 500, then the theoretical number of *different* DNA molecules or genes that could be formed would be 4^{500} (4 to the 500th power, or 4 multiplied by itself 500 times). This is an astronomically high number and thus fully accommodates the tremendous variety of genes that have ever existed, now exist, and will exist on the Earth. To emphasize this point in more familiar terms, we can use the same simple arithmetic in an analogous situation. If we have 5 pennies and arrange them in a row, how many different arrangements can we make? Each penny has 2 faces, heads and tails, so that we could have 5 heads, 5 tails, or various combinations of heads and tails in 30 different sequences. This is predicted from the value of 2^5 (2 multiplied by itself 5 times), which equals 32. The more pennies we include in our example, the more kinds of arrangements we can generate. If we used 10 pennies and arranged them in every possible sequence in rows of 10, then we could produce 2^{10}, or 1024 different combinations. You can see from these examples that doubling the number of pennies more than doubled the number of unique sequences; in fact, we have *squared* the variety since the product of $32 \times 32 = 1024$. These same arithmetical relationships were used by Malthus in his treatise on populations since population increase follows an *exponential* progression, whereas increases in food supplies and other essential resources tend to increase *linearly*. But, back to the genes. The exponential increase in gene variety that results from adding on more bases to make longer DNA molecules could easily explain how genes have kept pace with evolutionary changes in three billion years. At the same time, this feature of DNA provides one more astonishing virtue that helps us to understand its successful endurance since life originated on the Earth. Indeed, it is hard for us to envision a substitute chemical that could display the same virtuosity as the DNA molecules that function as genes.

This great variety of genes represents stored potential because the information is packaged within DNA molecules and is not immediately available to the organism in this miniaturized form. The information is coded, very much like the dots and dashes of the Morse Code or similar systems. The deciphering of the genetic code was a momentous achievement, which was completed with astonishing speed after the first real

Basic Genetic Mechanisms 15

breakthrough in 1961. At that time, it was shown that a particular combination of bases in nucleic acid could be translated into one of the 20 *amino acid* building blocks of proteins. It had been suggested earlier that triplets of bases which included various combinations of T, A, G, and C were the codewords for the amino acids, but there are 64 possible triplet codewords (4 bases in threes equals 4^3, or 64) and only 20 amino acids included in the code. Because of this discrepancy, we did not know how many of the possible triplets actually did specify particular amino acids, nor did we know whether more than one of the codewords could be translated into the same amino acid. Between 1961 and 1964, there was definitive evidence that was provided which clearly showed that 61 of the 64 possible triplets were codewords for amino acids and that from 1 to 6 different triplets might specify the same amino acid in the set of 20 (Fig. 1.5). A little later, the final evidence was presented which showed that the remaining three triplet combinations acted as "punctuations," such as those that signal the termination of a gene message. A gene thus includes a coded sequence of bases that is translated into a particular linear sequence of amino acids that comprises the protein product of gene action. The precise sequence of codewords in DNA determines the precise sequence of the amino acids in the protein coded by that gene, and the diversity of proteins is the result of the many kinds of genes made of the deceptively simple DNA molecule.

The genes are located within chromosomes in the nucleus of the cell while the proteins are manufactured in the cytoplasmic compartment surrounding the nucleus. How does the coded information in the nucleus get to the cytoplasm where translation into proteins takes place? It was also in 1961 that the experimental evidence was provided which showed that another kind of nucleic acid, ribonucleic acid or RNA, acted as a messenger that carried the coded information from DNA to the sites of protein synthesis in the cytoplasm. This messenger-RNA is transcribed faithfully from the DNA, and then the transcript messenger molecules travel to the cytoplasm where translation into proteins occurs (Fig. 1.6).

We might ask now why some kinds of proteins coded by our DNA are manufactured only in some of our cells and not in all of them. We do know that each cell in the body contains a complete and identical set of genes, yet we are an amalgam of different cells, tissues, and organs. The answer to this question is that genes are not active all of the time in all of the cells; that is, genes are turned off and on according to regulatory signals of various types. This phenomenon of **differential gene action** probably is universal, and it helps to explain why blood proteins are manufactured only in certain cells at certain times even though any cell has the gene potential to specify these proteins. In fact, there is good

Second nucleotide

First nucleotide	A or U		G or C		T or A		C or G		Third nucleotide
A or U	**AAA** UUU / **AAG** UUC	Phe	**AGA** UCU / **AGG** UCC	Ser	**ATA** AUU / **ATG** AUG	Tyr	**ACA** UGU / **ACG** UGC	Cys	**A** or U / **G** or C
	AAT UUA / **AAC** UUG	Leu	**AGT** UCA / **AGC** UCG		**ATT** AUU / **ATC** AUC	Stop	**ACT** UGA / **ACC** UGG	Stop / Trp	**T** or A / **C** or G
G or C	**GAA** CUU / **GAG** CUC	Leu	**GGA** CCU / **GGG** CCC	Pro	**GTA** CAU / **GTG** CAC	His	**GCA** CGU / **GCG** CGC	Arg	**A** or U / **G** or C
	GAT CUA / **GAC** CUG		**GCT** CCA / **GCC** CCG		**GTT** CAA / **GTC** CAG	Gln	**GCT** CGA / **GCC** CGG		**T** or A / **C** or G
T or A	**TAA** AUU / **TAG** AUC	Ile	**TGA** ACU / **TGG** ACC	Thr	**TTA** AAU / **TTG** AAC	Asn	**TCA** AGU / **TCG** AGC	Ser	**A** or U / **G** or C
	TAT AUA / **TAC** AUG	Met	**TGT** ACA / **TGC** ACG		**TTT** AAA / **TTC** AAG	Lys	**TCT** AGA / **TCC** AGG	Arg	**T** or A / **C** or G
C or G	**CAA** GUU / **CAG** GUC	Val	**CGA** GCU / **CGG** GCC	Ala	**CTA** GAU / **CTG** GAC	Asp	**CCA** GGU / **CCG** GGC	Gly	**A** or U / **G** or C
	CAT GUA / **CAC** GUG		**CGT** GCA / **CGC** GCG		**CTT** GAA / **CTC** GAG	Glu	**CCT** GGA / **CCC** GGG		**T** or A / **C** or G

figure 1.5

The genetic code. The base triplets in DNA are shown in boldface; the RNA codewords of the messenger molecules are complementary to the codewords in DNA. Except for three "punctuation" codewords that signal the end of a genetic message, the remaining 61 combinations are translated into the 20 amino acids that are typical of most proteins. Some amino acids are specified by only one codeword (methionine, tryptophan), and others by as many as six different triplets (leucine, serine, arginine). Note that the distribution of codewords that specify the same amino acids is not random, probably reflecting evolutionary relationships. (The DNA codons appear in boldface type; the complementary RNA codons are in italics: A=adenine; C=cytosine; G=guanine; T=thymine; U=uridine (replaces thymine in RNA). In RNA, adenine is complementary to thymine of DNA; uridine is complementary to adenine of DNA; cytosine is complementary to guanine, and vice versa. "Stop"=punctuation. The amino acids are abbreviated as follows: Ala, alanine; Arg, arginine; Asp, aspartic acid; Asn, asparagine; Cys, cysteine; Gln, glutamine; Glu, glutamic acid; Gly, glycine; His, histidine; Ile, isoleucine; Leu, leucine; Lys, lysine; Met, methionine; Phe, phenylalanine; Pro, proline; Ser, serine; Thr, threonine; Try, tryptophane; Tyr, tyrosine; Val, valine.)

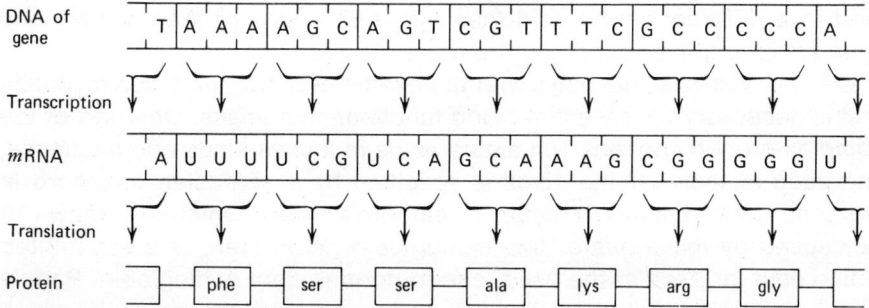

figure 1.6
The information contained in the sequences of coded DNA are *transcribed* to molecules of messenger RNA, which travel to the cytoplasm outside the nucleus. The *translation* of the codewords into amino acids of the proteins takes place in the cytoplasm, according to the original instructions in the DNA blueprints.

evidence that as many as 95 percent of the total number of genes in any one of our cells may be turned off most of the time. But, different genes are active in different cells so that many different cell types are produced that can perform many different functions. Differential gene action also explains the sequences of development and differentiation that occur during the lifetime of an individual; some of the genes are active in some of the cells some of the time. If all or most of our genes perform properly at the appropriate times, then a normal functioning individual is produced. Because of the complexities of timing and interactions, however, some people suffer the consequences of improper or defective gene action. Not all malfunction or malformation is caused by genes; many deficiencies and defects in humans and other organisms result directly from nongenetic causes or from some combination of genetic and environmental influences.

Mutation

A sudden heritable change is known as a **mutation.** Because the link between generations resides in the genes transmitted from parents to progeny and because genes are made of DNA, a mutation must involve an alteration in the DNA itself. Mutant genes would be transmitted in the same way as nonmutant DNA, but their translations (proteins) would be different. We know of various kinds of changes in chromosomes and genes that can be called mutations, but the simplest mutational event is one in which a single codeword is altered so that a single amino acid of a large protein molecule is different in the mutant and nonmutant individuals. In fact, a change in only one of the three bases of a codeword could lead to a different amino acid (see Fig. 1.5), and a change in only one amino acid might render the entire protein defective.

We can illustrate this point with the example of the hemoglobin protein that is necessary for essential blood functions in humans. One part of the complex protein contains 104 amino acids in known kinds and sequence, and each of these amino acids is specified by a particular codeword in the gene DNA. Among a number of blood disorders, which are known to be caused by mutations of this hemoglobin gene, there is a substituted amino acid in place of the usual one in normal adult hemoglobin. People who have sickle cell anemia produce hemoglobin molecules in which amino acid number six is *valine* instead of *glutamic acid*. A change in the middle base of the triplet DNA codeword from *thymine* to *adenine* may be the only result of the gene mutation (Fig. 1.7). But, this one change leads to the defective hemoglobin protein which causes the pain, crises, and shortened life of people who develop the sickling disease.

All mutations do not cause disease and distress, as we can see clearly from the ever-improving progression of life forms during evolution. New characteristics and new species develop as a result of successful mutational changes, and diversity within a species also attests to the incorporation of mutations into the gene pool of species populations. Different human races share a common set of genes, but there are enough genetic differences for some groups to be distinguished from others. These differences are not due to different genes, but rather to different mutational alternatives of the same genes. There are a number of genes that determine skin color, but members of the different races have differences in one or more codewords of the same genes and thus produce slightly different pigments in the skin cells. People who have brown eyes and people who have blue eyes have the same gene for eye color, but the DNA molecule is slightly different in each case, and the pigment proteins reflect this difference. The accumulation of mutational changes in the genes leads to the steady changes in all life forms and to the procession of species that have inhabited this world for three billion years.

Mitosis and Meiosis

The two major processes by which sets of chromosomes are distributed to the daughter cells are **mitosis** and **meiosis**. Mitosis occurs in many kinds of cells and leads to two daughter nuclei that are exactly like the mother nucleus from which they were derived. Meiosis is a nuclear division process which occurs only in special kinds of cells, often only during certain periods of the life cycle, and it is an essential component of sexual reproduction.

figure 1.7
The mutation in the gene that directs the manufacture of the hemoglobin protein may occur in various places within this gene. Two of the mutations happen to affect the same amino acid position (number 6) in the 104 that constitute this molecule. It is probable that a change in one base of the triplet codeword could have led to each of these particular amino acid replacements in the altered form of the molecule.

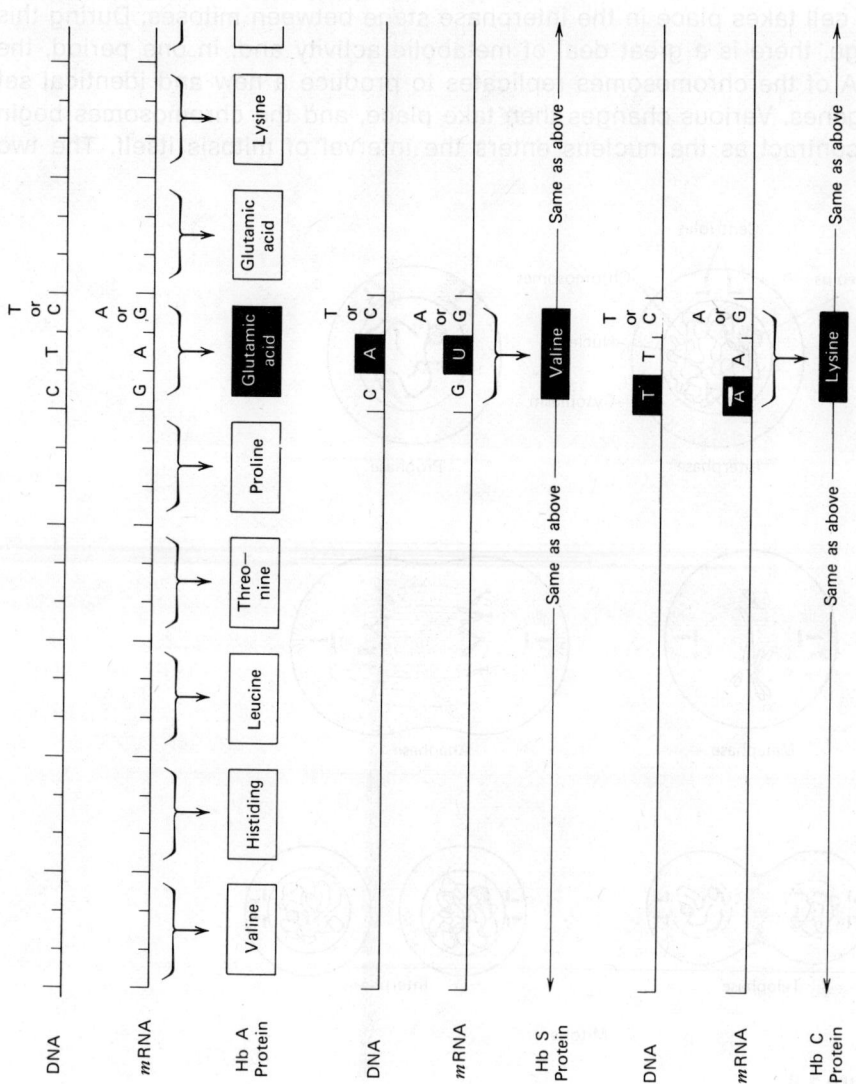

Mitosis

The mitotic division of a nucleus to produce two identical daughter nuclei requires a prior duplication of the chromosomes (Fig. 1.8). If we were to examine cells under the microscope during the time that DNA was replicating, we would be unable to see the process, and we could not even distinguish the separate chromosomes. The most active time in the life of the cell takes place in the interphase stage between mitoses. During this stage, there is a great deal of metabolic activity and, in one period, the DNA of the chromosomes replicates to produce a new and identical set of genes. Various changes then take place, and the chromosomes begin to contract as the nucleus enters the interval of mitosis itself. The two

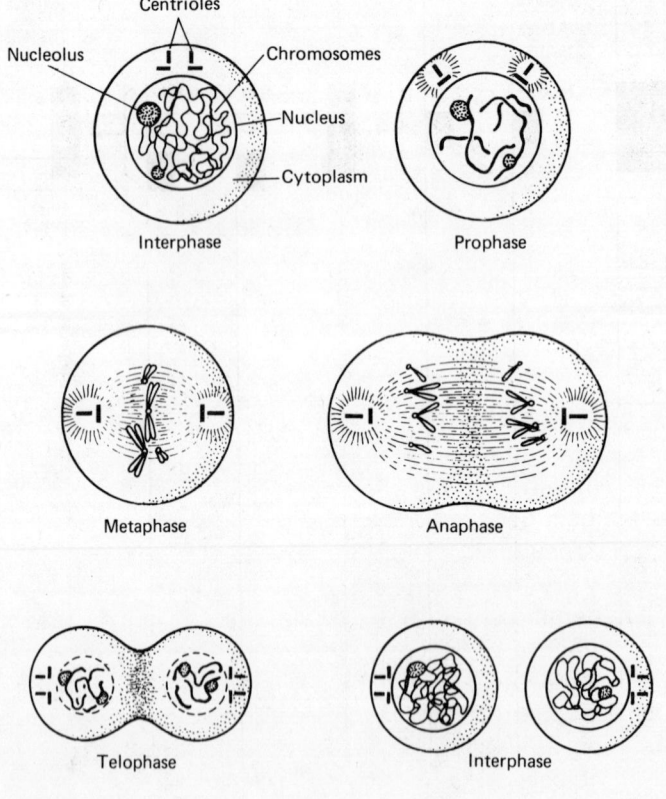

figure 1.8
The sequence of events in mitosis is subdivided arbitrarily into particular stages for convenience of reference. Actually, the process is continuous and there are no stopping points during the mitotic division of a nucleus in a cell.

sets of chromosomes are separated as the old and the new structures are transported to opposite poles of the cell via special fibers. Then the two identical nuclei reorganize and finally resume the interphase stage, which signals the completion of the division process. The daughter nuclei usually are segregated into separate cells by a cell division process that is coordinated with mitosis, but which is not an essential partner for nuclear division. It is the process of mitosis that is responsible for the increase from the one cell of the fertilized egg, which is the beginning of us all, to the trillions of cells that are produced afterward. The daily replenishment of our body cells is possible because some of our tissues retain the capacity for mitosis even while others lose the ability to multiply.

Mitosis was one of the most significant evolutionary inventions because the same mechanism serves equally well for a nucleus with two chromosomes as for others which have two hundred. Since this distribution process is accurate for cells regardless of chromosome number, increases and decreases in the number of chromosomes during evolution do not impose restrictions on species success. The particular number of chromosomes that characterizes a species is not an indication of evolutionary rank. There are 46 chromosomes in human body cells, 48 in the chimpanzee, 14 in the garden pea studied by Mendel, and a great range of numbers throughout the biological kingdoms. Mitosis occurs in essentially the same way in all of these species, which is further evidence of the continuity of life forms during evolution.

Meiosis

Asexual systems lack meiotic capacity. Sexual systems undergo meiosis, but the particular kind of cell and the precise time of the meiotic nuclear divisions varies among different species and even between the two sexes of the same species. Meiosis generally occurs only in the sex organs, or **gonads**, in animals, and results in the production of the sex cells, or **gametes**. Gametes usually are the sperm and eggs if their size and behavior warrant this specific identification. Any cells that are capable of fusion to produce a new individual are gametes, whether they look alike or not, and whether or not their behavior is the same.

The situation is a little different in most plants because there is a phase of spore production as well as gamete production, but the main events are still those of meiosis and fertilization. The products of meiosis are cells that have half the number of chromosomes as the mother cell (Fig. 1.9). If there was no process by which the chromosome number could be reduced to one-half, then there would be a doubling of numbers in every generation because the gametes fuse together to form one cell containing one nucleus. In a very short time, a species would have accu-

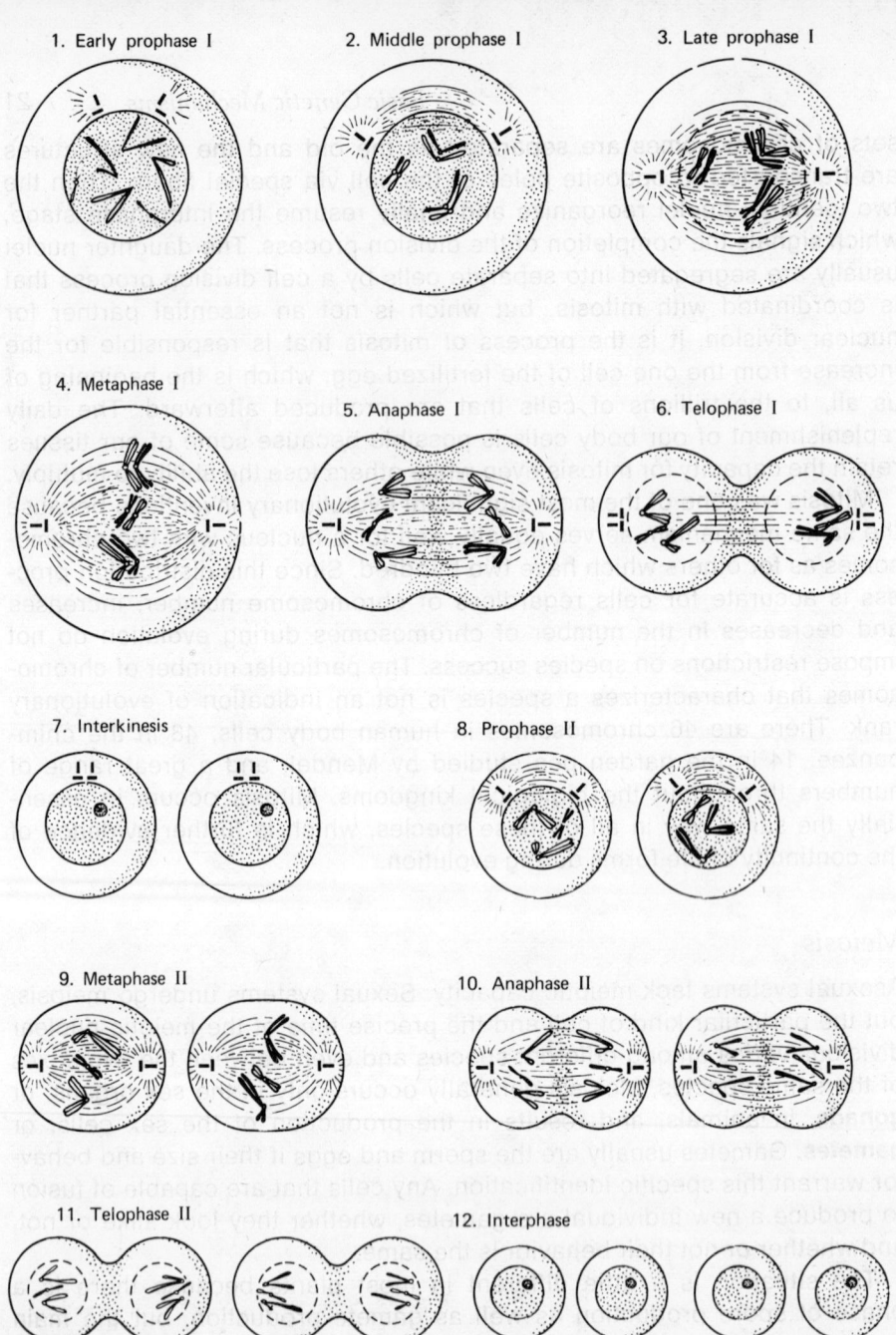

figure 1.9
The meiotic nuclear divisions lead to halving of the chromosome number in the final products. Since DNA duplicated before the first division began, each chromosome is a double structure and remains so until the second division when single chromosomes are distributed to opposite ends of the cell. Four products generally are formed, but three of these degenerate in most cases of oogenesis, whereas all four survive in spermatogenesis.

mulated millions of chromosomes, and we know that this does not happen. In fact, it was predicted that some process would be found to explain the constancy of chromosome number for a species. After the discovery of fertilization in microscope studies reported in 1875, it was clear that the doubling of nuclear content must have a compensatory partner process in which the chromosome number would be reduced by one-half in the gametes. Meiosis was not reported for some years later, but once the evidence became available, it fit right in with the predictions that had been made.

Every individual of a species has the same chromosome number (some exceptions always occur), and this number is maintained in generation after generation. In humans, there are 46 chromosomes in the body cells, one set of 23 having come through the egg and the other matching set from the sperm. Fusion of the egg and sperm results in a new cell that has 46 chromosomes. Repeated mitotic divisions of this fusion nucleus and its descendant nuclei are responsible for the development of the new human being whose cells each contain 46 chromosomes. The egg and the sperm cells are products of meiosis, and each therefore contains only half of this number, or 23 chromosomes. Cycles of reproduction include meiosis and fertilization in sexual species, from the simplest one-celled forms to the human species.

Formation of Gametes

By and large there is a morphological difference between eggs produced by females and sperm produced by males. In the case of birds, the egg is an enormously enlarged cell that is many millions of times bigger than the microscopically sized sperm. In humans and most other mammals, the gametes of both sexes are microscopic in size; but the egg is larger than the sperm and also different in shape, nutrient content, and behavior. Except for the sex chromosomes, the egg and sperm are equivalent in gene content so that there is essentially an equal contribution from each parent to the next generation.

The **oocytes** are specialized cells in the ovary, each of which undergoes meiosis to produce one functional egg and three abortive cells that are discarded (Fig. 1.10). The egg cell is rich in nutrients and in cellular structures that are important in the manufacture of chemicals and in the generation of energy by which these chemical reactions can proceed. The sperm contain very little nutritive material, but there is an energy-generating system that aids in sperm movement and, most importantly, a set of genes within the chromosomes of the sperm nucleus. One sperm fertilizes the egg, and all of the others that swarm around the egg are prevented from entering because of the rapid modifications that make the outer covering of the newly fertilized egg impervious to other sperm.

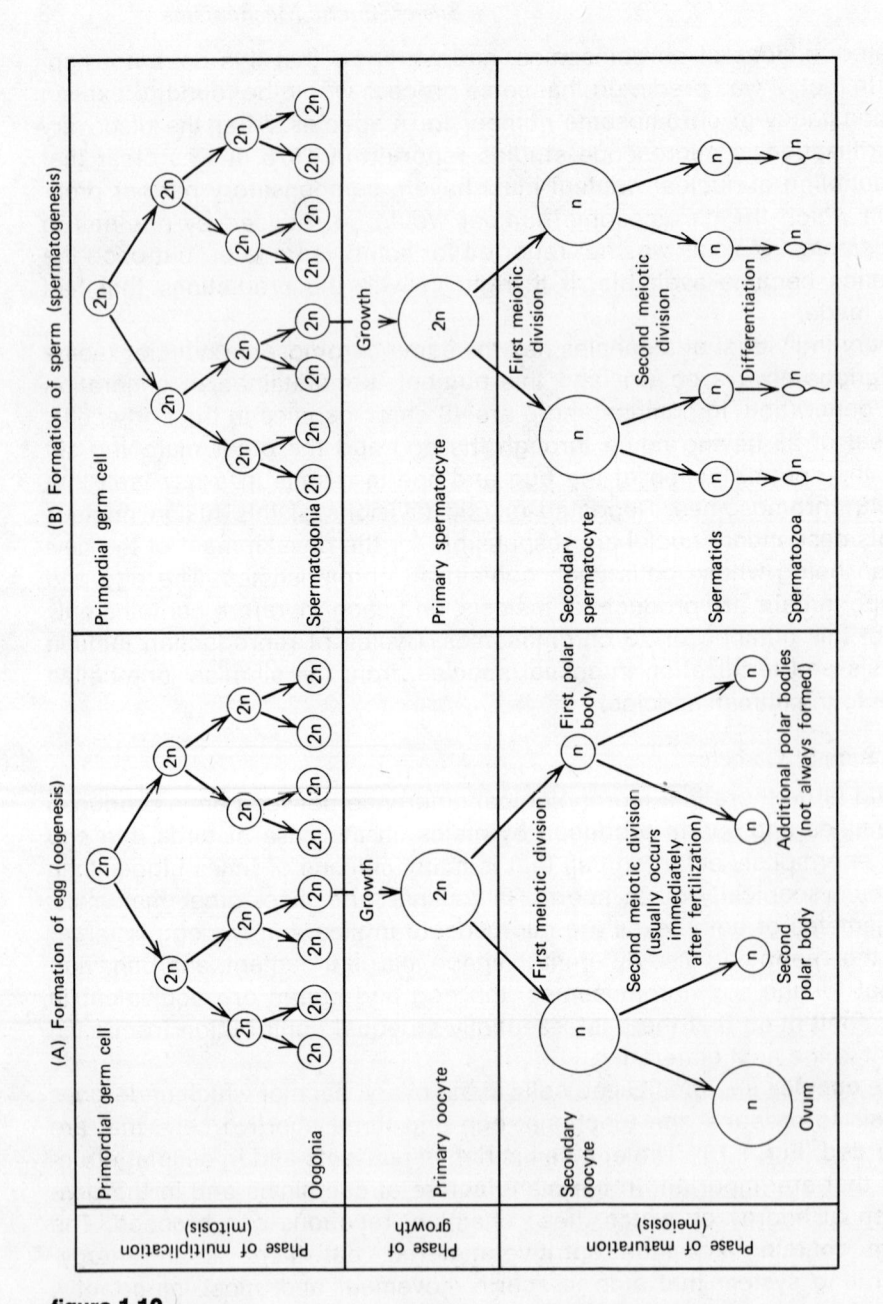

figure 1.10

In higher animals and in plants, oogenesis leads to one functional egg, whereas spermatogenesis typically results in four functional sperm at the conclusion of the meiotic divisions. The significant feature is the same however; the gametes have half the number of chromosomes as the parent cell.

Once the egg and sperm nuclei have fused, the cell is a new individual with two sets of chromosomes and genes that are derived from both parents.

The fertilized egg develops into an embryo and ultimately into a new individual which is recognizably like its parents in most features. The sequence of embryonic development to the adult varies among different animals. Some insects and some vertebrates such as frogs, among others, go through a developmental program that includes a larval stage, metamorphosis, and then they finally emerge in a form that resembles the adult of the species. For many other animal species, the young are born resembling the adult, except for immature and undeveloped features that change with time as the individual matures to reproductive capacity.

The sperm-producing organs generally produce gametes either continually or during breeding seasons, once the male is mature. Four functional sperm are produced from each **spermatocyte** as a result of meiosis (Fig. 1.10). In humans, the sperm are produced and mature to become motile **spermatozoa** continually between puberty and old age; some hundreds of millions of sperm are produced every day. Women are born with about one million oocytes already at an early stage of meiosis, and no additional oocytes are produced for the rest of her life. The oocytes remain inactive until puberty, at which time meiosis is resumed in one or more of the oocytes and one or more eggs, or **ova**, are produced each month. This process continues until the woman is in her 40s or 50s. The end of ovum production marks the beginning of menopause during which there are particular hormonal changes that occur, as well as the termination of fertility.

Failure to produce gametes does not influence sexual activity in humans; however, it makes it impossible to sire or bear children. The phenomenon of sex in humans includes gamete production and the property of reproduction, but many other physiological and behavioral features also contribute substantially to the expressions of sexual activity. The ability to procreate is separable from other aspects of human sexuality, and this separability constitutes one of the unique and constant features of the human species. In most animal species, the sexual acts are pursued at times that are most likely to lead to the production of offspring, but humans are not restricted biologically, psychologically, or socially to this pattern. We will discuss these qualities in later chapters.

The Determination and Differentiation of Sex

In mammals and many other animals, body cells of females carry two **X** chromosomes and two sets of autosomes, while male cells are **XY** and carry the same two sets of autosomes. During meiosis females produce ova, each of which contains one set of autosomes and one **X** chromosome. Males produce two kinds of sperm, sperm that have an **X** and sperm that have a **Y** chromosome, along with an autosome set. Since any egg may be fertilized by any sperm, fusions of gametes are random and the combinations are such that half of the eggs are fertilized by sperm that have a **Y** chromosome and half by **X**-carrying sperm. Stated another way, we would say that the chance of an egg being fertilized by a sperm with an **X** chromosome is one-half, and the chance of fertilization by a **Y**-bearing sperm also would be one-half. Since each egg has an **X** chromosome and since fertilization is a random event, half of the resulting population will be **XX** (female) and half of the population will be **XY** (male). This sex ratio of 1:1 is an outcome of the randomness of fertilization and the occurrence of two kinds of sperm but only one kind of egg in relation to their sex chromosome. Exactly the same kind of ratio can be obtained if we consider an analogous situation involving two coins. If we place one coin on the table head side up and we toss the second coin, how many times will the second coin land on heads and how many times on tails? If there are enough tosses to make a reasonable sample, we would find what you already have predicted: half the time the combination of the two coins would be heads–heads, and half the time it would be heads–tails, purely due to chance.

The moment of gamete fusion is the moment of the **determination** of sex since that is the time when either the **XX** or the **XY** combination is achieved. The fertilized egg that is **XX** cannot be distinguished from one that is **XY**, but each has a different potential for development. The **differentiation** of sexual characteristics takes place during the period of development in the embryo and continues until maturation to adulthood, guided and modulated by the genetic and physiological properties of the individual.

The **primary sex characteristics** are the particular gonads that are present and functioning, so that females have ovaries and males have testes (or organ equivalents in the great diversity of life forms). The human embryo contains one pair of undifferentiated gonads until six to twelve weeks of age. At about six weeks after conception, an embryo that is **XY** begins to differentiate its primary sex characteristic of testis development from the inner tissues of the primordial gonads, whereas an **XX** embryo undergoes ovarian differentiation from the outer layers of these same gonads sometime around the twelfth week after conception. Gonad differentiation as ovary or testis depends on the presence of a **Y** chromosome for males and two **X** chromosomes for female embryos. If a **Y** chromosome is present, even though extra **X** chromosomes also occur in aberrant individuals, testis differentiation will begin early in embryonic existence. If there is no **Y** chromosome, then differentiation is delayed and, when the gonad does develop, it becomes an ovary in female embryos that have at least two **X** chromosomes. In those occasional embryos that have only one **X** and no **Y** chromosome, female differentiation occurs, but the gonad remains undeveloped and never functions in the adult. As far as we know, the sex chromosomes exert their effects only during this initial phase of sex differentiation and probably do not influence major processes of sexual development afterward, except for certain genes about which we know very little.

Once the gonads have differentiated they begin to secrete sex hormones, which then influence the differentiation of the internal reproductive structures and finally the external genitalia. A three-month-old fetus can be recognized as a male or a female by the external genitalia, but this is not possible at an earlier stage. The influences of sex hormones that are produced in the gonads and other glands continue throughout a major portion of the life of the individual, especially in the differentiation of secondary sex characteristics. Any feature that is different in males and females, except for the gonads themselves, can be considered a **secondary sex characteristic.** Examples of these would include differences in patterns of hair growth, voice pitch, breast development, amount and distribution of muscle and fat, and other features in humans; the size of the adult, prominence of the canine teeth, horn or antler development, and many other physical traits by which we recognize males and females of many forms of life. In most cases we need only a single external clue to label an animal as male or female, and it need not be the genitalia. The lion's mane, the brilliant red plumage of the cardinal, or the large canine teeth of the baboon are sufficient to identify the male; similar features distinguish the females, although more often it is a matter of the female *not* showing particular traits (in the curious ways that descriptions tend to be written).

There are, of course, many variations on the basic theme. Most plants and some animals are **hermaphroditic** species in which each individual can function both as male and female during sexual reproduction. Such species as the earthworm among animals, and corn as an example among plants, rarely self-fertilize; that is, the gametes are usually exchanged between different individuals so that one organism both receives and delivers sperm during a copulatory or fertilization event. Many flowering plants are self-fertilizing, but this is not true for animals.

Whatever the variation in organ location or formation, a species has the capacity for sexual reproduction if gametes are produced and can fuse to initiate the new generation. The sexual life cycle includes the punctuating processes of meiosis and fertilization, which not only permit the constancy of chromosome number but which also lead to possibilities for genetic diversity that cannot be realized by asexually reproducing species.

Suggested Readings

Beermann, W., and U. Clever, Chromosome puffs. *Scientific American*, 214 (April 1964), 50.

DuPraw, E. J., *DNA and Chromosomes*. New York: Holt, Rinehart, Winston, 1970.

Ingram, V. M., How do genes act? *Scientific American*, 198 (January 1958), 68.

Loewy, A. G., and P. Siekevitz, *Cell Structure and Function*, 2nd ed. New York: Holt, Rinehart, Winston, 1970.

Nirenberg, M. W., The genetic code: II. *Scientific American*, 212 (March 1963), 80.

Watson, J. D., *Molecular Biology of the Gene*, 2nd ed. New York: W. A. Benjamin, 1970.

Yanofsky, C., Gene structure and protein structure. *Scientific American*, 216 (May 1967), 80.

chapter two

Sex-Related Inheritance Patterns

Since the genes are located in chromosomes, we would expect that the inheritance of genes would vary in pattern depending on whether they were part of an autosome or a sex chromosome, and they do. In addition to these two kinds of inheritance patterns, there are others that show differences in relation to sex, but not in relation to the chromosomal location of the gene itself. The general principles that have been formulated from studies of known genes can be used to analyze and interpret the inheritance of new genes at any time that investigations can be performed. At the same time, we are able to predict the proportions of families or populations who will display a genetic trait once we know the particular kind of inheritance pattern that is assigned for it.

Patterns of Inheritance

Most of our inherited traits result from the actions of many genes, but the analyses are quite complex. The principles of inheritance can be illustrated equally well by using simpler one-gene examples, and we will refer only to these patterns in the discussion that follows.

Autosomal Inheritance

The genes in autosomes are equivalent in the two sexes so that we would expect autosomal genes to be distributed equally to males and females in a progeny. This prediction can be verified by examining the progenies from reciprocal matings; that is, by determining the distribution of each gene alternative when it is transmitted through the female parent in one set of matings and through the male parent in reciprocal matings. Of course, we would be unable to conduct a genetic study without knowing at least two alternative forms of the same gene; such alternatives are called **alleles** of a gene. One example would be the gene that is responsible for the manufacture of an amino acid that is necessary for the synthesis of one of the pigment proteins in skin, eyes, and hair. We all have this gene and most of us have the particular allele which is functional, so that the pigment is produced. Some people are *albino* and lack this pigment. Such an albino has the same gene, but in the alternative form so that the allele in the albino is incapable of directing the production of the needed amino acid; hence pigment is not manufactured and such individuals have hair and skin that are virtually colorless and eye color that is very pale or almost pink because of seeing right through to the blood capillaries in the eyes. Many of the familiar white rabbits and rats that we use in the laboratory are albinos, too.

If we were to conduct genetic analyses using mice, fruit flies, or corn plants, all of which are favorite experimental species, we would set up our experiments using stocks of pure-breeding individuals. With mice from a gray strain and mice from an albino strain, we would arrange matings reciprocally so that gray males were mated with white females and white males with gray females. These parental mice breed true within their own strains, grays producing only grays and whites producing only whites, but what happens when mice from the different strains are interbred? When we examine the progenies from each of these sets of reciprocal matings, we find that all the young mice are gray and none of them are albino. This information by itself is incomplete since similar results could occur for various genetic reasons. We must proceed further and interbreed this first generation, or F_1, making sure to keep the two set of F_1 progenies separate in the matings so that interbreeding is permitted within each of the reciprocal progenies, and not between them. The second or F_2 generation of mice are quite different from their F_1 parents. In each of the F_2 progenies, we would find mice that were gray and others that were albino. Furthermore, we would count the actual numbers of grays and whites to see if there is the predicted ratio for a case of autosomal inheritance of a single pair of alleles for one gene. Since the time of Gregor Mendel, we have known that a 3:1 ratio is produced in such inheritance patterns. When we make the counts and convert these numbers to proportions of grays to whites, we find that 3/4 of each F_2 progeny are gray and 1/4 are albino. This set of results is expected theoretically for single-gene autosomal inheritance (Fig. 2.1).

These results illustrate some fundamental principles of inheritance. The allele for gray fur is **dominant** and the allele for albino is **recessive** to it. Since each parent contributes one set of genes through the gametes, each of the F_1 mice must have received one allele for gray and one for albino coloration. Since the albino allele was masked from expression when in the same nucleus with the allele for gray, all the F_1 animals developed gray fur. We know that the two kinds of alleles were present in the F_1 mice because the F_2 progeny included albino as well as gray types. The proportion of 3/4 gray to 1/4 albino, or 3:1, in the F_2 generation, is predictable when we know that each gamete has only one allele of every kind in the species and that gamete fusion is random. Fusions of eggs and sperms in all possible combinations on a random basis leads to 1/4 that have only alleles for gray, 1/2 that have one gray and one albino allele, and 1/4 with only albino alleles. This last group is the only one that develops the albino characteristic since all the others have at least one allele for gray, which is dominant. Precisely this same proportion is obtained if we use our familiar coin analogy. If we toss two coins one hundred times (a reasonable sampling), we would find that 1/4 of the time it would be heads–heads, 1/2 the time heads–tails, and 1/4 of the time tails–tails; due to chance alone.

The F_2 progenies from each set of matings produced the same results so that we know an autosomal gene was responsible for the fur color trait. It made no difference whether the albino parent was a male or a female, which means that there was an equal genetic contribution by each parent to the progenies, in each generation. This indicates clearly that the gene is in an autosome since eggs and sperm have similar sets of these chromosomes. The adult mice have two sets of autosomes in their body cells, but after meiosis the gametes have only one set. When the gametes fuse at fertilization, the new generation begins with two chromosome sets. It should be clear that the rearrangements of alleles and the particular proportions of the progeny types are a reflection of the **segregation** of alleles during meiosis and their **recombination** at the time of fertilization. Sexual reproduction permits gene recombinations and adds to the diversity of populations.

Sex-Linked Inheritance

The first mutant of the common fruit fly, which was analyzed in the famous laboratory of Thomas Hunt Morgan at Columbia University, was a white-

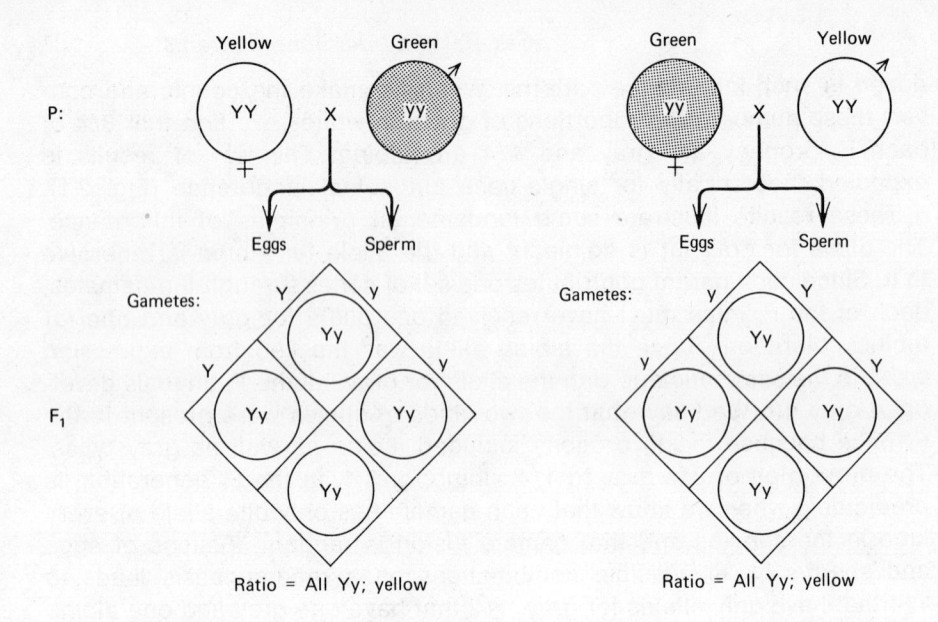

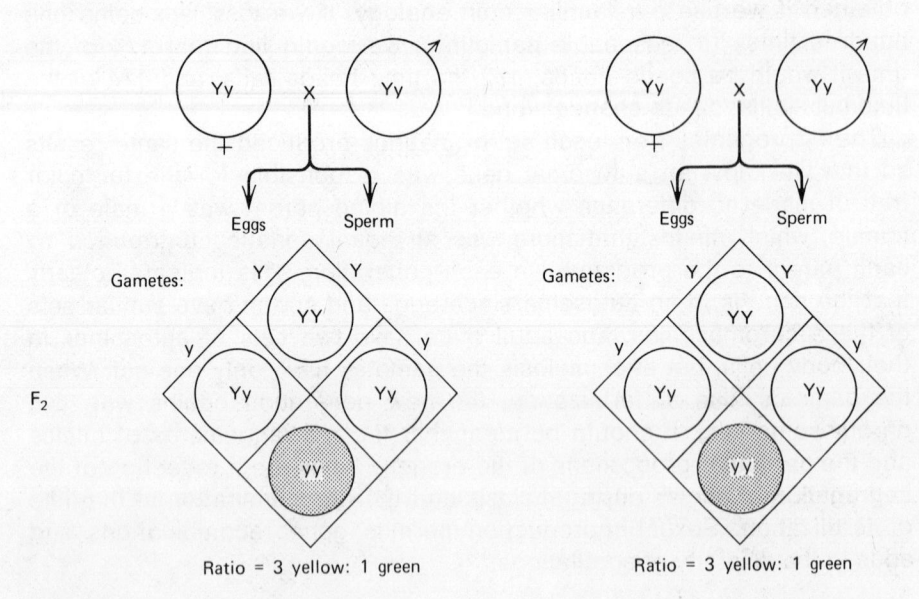

figure 2.1
Reciprocal matings produce identical progenies in autosomal gene inheritance. The trait shows the typical 3:1 ratio of classes in the F₂ generation. The allele that is dominant is the one that is expressed in the F₁ generation and in three-fourths of the F₂.

eyed form. A male spontaneous mutant was discovered in their stocks of the usual red-eyed flies, and he was bred with red-eyed females. All the F_1 progenies had red eyes, which tentatively indicated that red was dominant to white eyes. When these F_1 flies were interbred to produce the F_2 generation, 3/4 had red eyes and 1/4 of the F_2 were white-eyed. This confirmed the dominance of red to white, and also showed that one pair of alleles of a single gene was responsible for the eye color difference. But there was one striking variation in this F_2 progeny from the expected; namely, that all of the females had red eyes but half of the males were red-eyed and half had white eyes. The F_2 ratio still was 3:1, but the two classes of eye color were not distributed equally between the sexes. Subsequent breeding experiments were performed using pure-breeding white and red-eyed flies in reciprocal matings. These results provided further evidence that the distribution of the alleles was linked to the sex of the parent that contributed the allele to the progenies (Fig. 2.2). If you examine these results you can see that reciprocal progenies are different both in the F_1 and the F_2 generations. One consistent feature of sex-linked inheritance is this difference between reciprocal progenies, whereas autosomal inheritance yields identical reciprocal progenies in the F_1 and F_2 generations.

Sex-linked inheritance patterns reflect a distribution of alleles that is coincident with the numbers and distribution of the **X** chromosome in males and females, so that it sometimes is called **X**-linked inheritance. Females produce only one kind of egg, all of which contain one **X** chromosome; males produce two kinds of sperm, half bearing an **X** and half with a **Y** chromosome. Since there are no alleles on the **Y** for those on the **X**, females may have any of the three combinations of the two alleles for the gene (two dominants, two recessives, or one of each allele), whereas males have either the dominant or the recessive allele in their single **X** chromosome. These features lead to some characteristic qualities of sex-linked inheritance that are not found in autosomal patterns. Sons receive their **X** chromosome through the egg and their **Y** chromosome through the sperm so that sons inherit their sex-linked genes only from their mother and not from their father. Mothers who have a different allele of a particular gene on each of their two **X** chromosomes are *carriers* of recessive traits but do not express them because the dominant allele on the other **X** chromosome masks recessive allele action. Their sons, however, can be of either the dominant or the recessive type, depending on which of the **X** chromosomes happened to be in the egg that was fertilized by a **Y**-carrying sperm that produced the **XY** male offspring.

Fathers transmit their sex-linked traits only to their daughters and not to their sons, in accord with their contribution of either an **X**- or a

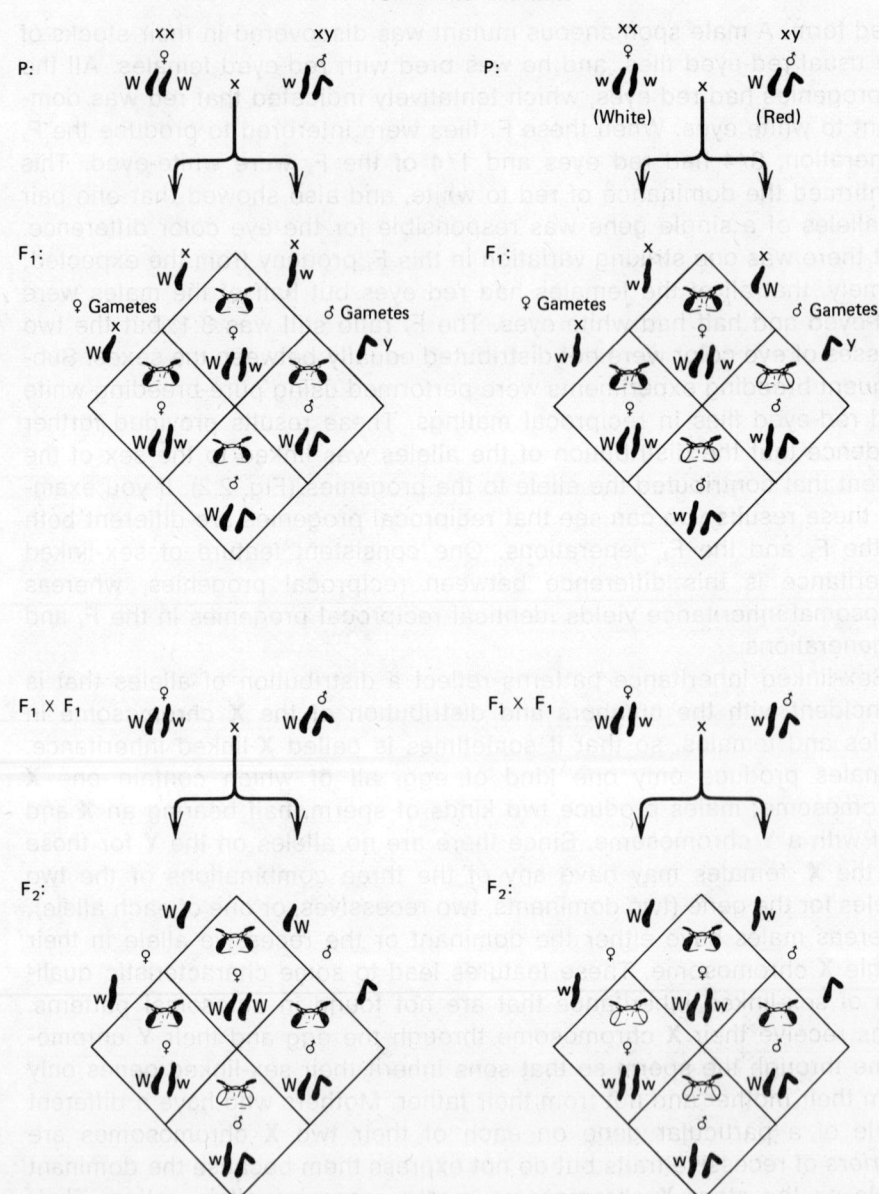

figure 2.2
The inheritance of white eye color follows a sex-linked inheritance pattern; the distribution of the alleles parallels the distribution of the X chromosomes and their number in the individual. Reciprocal matings lead to different proportions of the two classes in the progenies, but the pattern still reflects single-gene inheritance.

Y-carrying sperm at fertilization. Men who have the sex-linked trait of red-green color blindness will not produce color-blind sons. But they can transmit this allele to their daughters, who then would be carriers of the trait if they received the dominant allele on the **X** chromosome from the egg. This "criss-cross" pattern of transmission from mother to son, and from him to his daughters, is typical for sex-linked inheritance.

More males than females will express a recessive sex-linked trait since there is no second allele to mask the one derived through the egg. In many cases, it is extremely rare to find a female who expresses a recessive sex-linked trait since she would have to have been born of a mother who was a carrier and a father who carried and displayed the trait. Many known sex-linked traits are recessive and rare, so that the chances of two such people coming together from the general population would be quite low.

As we saw from the discussion of Morgan's white-eyed fruit flies, sex-linked inheritance produces different proportions of the two classes in each generation when the matings have been made reciprocally. These and other studies were conducted in the early decades of this century and, because of their enormous contribution to the new field of genetics, T. H. Morgan was awarded the first Nobel prize ever to be given to a geneticist. Many other geneticists have been made Nobel laureates since 1933, but most of the awards were made after 1960.

Other Inheritance Patterns Related to Sex

Sex-Influenced Inheritance

When the inheritance pattern does not conform to the distributions expected for either autosomal or sex-linked genes according to the conventional schemes we have just described, then we must look for the reason that underlies the variation. Even though the gene occurs in an autosome, as we can tell if reciprocal progenies are identical, the expression of the gene may be influenced by sex hormones. A classical example of such a sex-influenced autosomal pattern of inheritance is associated with a form of baldness in humans (Fig. 2.3).

From family histories we can see that the two alleles of the gene behave differently in males and females. Men will develop baldness when they have two of the possible three combinations of the alleles, but women will develop this trait only if one of these three allele combinations is present. In classical terms, the baldness allele behaves as a dominant in males and as a recessive in females. This allele is expressed in males whether or not the nonbaldness allele also is present, like a typical dominant, but it is expressed in females only when two baldness alleles occur

GENETICS AND EVOLUTION

Sex—influenced autosomal inheritance

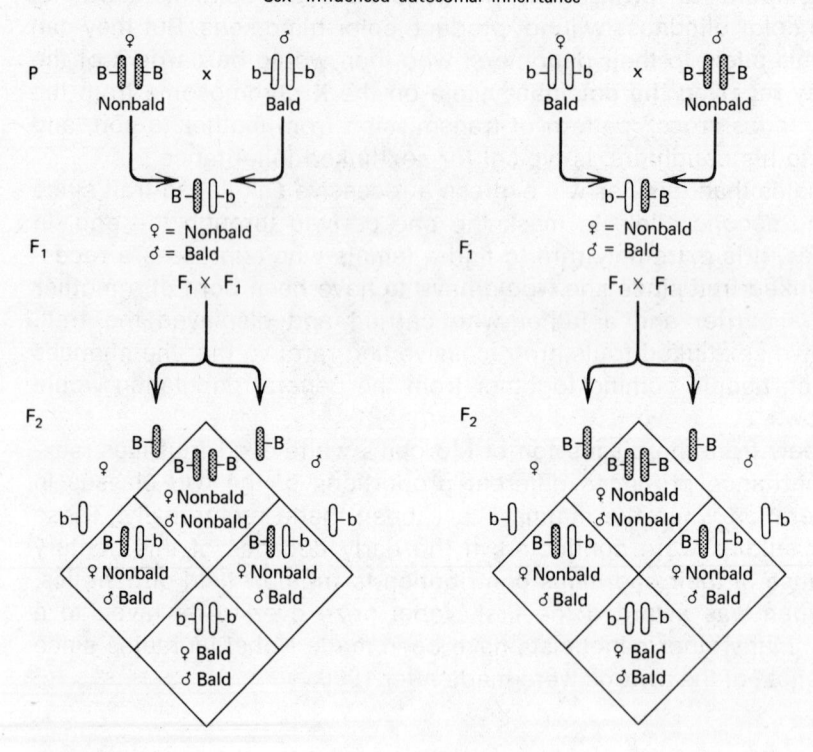

figure 2.3
Sex-influenced inheritance of pattern baldness in humans is autosomal (reciprocal progenies are similar) and based on one pair of alleles of a gene (F_2 ratio is 3:1), but the behavior of each allele as dominant or recessive is different in males and females of identical gene constitution. This trait is dominant in males (3:1 F_2 ratio) and recessive in females (1:3 F_2 ratio).

in the same cell, like a typical recessive. The proportion of the two hair patterns among men and women shows the expected 3:1 ratio of single-gene autosomal inheritance, but the behavior of the allele as a dominant or recessive is influenced by the predominant sex hormones of the individual.

The degree of baldness that develops also is modulated by the sex hormones so that a less drastic loss of hair occurs in women than in men, even when they have the same combination of alleles for this trait. Castrated males (testes removed) behave genetically like females in the development of this form of baldness, which further substantiates the relationship of gene expression to the influence of sex hormone. The level of secretion of male sex hormone probably is the major factor that regulates the action of the baldness allele. Although females and castrated males produce some amount of the weaker male sex hormones, they produce none of the potent hormone testosterone and are less likely to express baldness in this particular genetic situation.

Sex-Limited Inheritance

In this variation on the theme of autosomal inheritance, one of the two alleles of a gene never (or rarely) is expressed in one of the sexes but shows typical dominance/recessiveness behavior in the other sex. Sex-limited traits generally involve the primary or secondary sex characteristics of the individual. Genes for egg production occur in both sexes, but they are only expressed in the egg-producing females. Progenies of reciprocal matings are identical, thus indicating autosomal inheritance, but only one of the sexes will display the typical 3:1 ratio even though both sexes possess all of the possible allele combinations.

Both men and women have genes that determine the development and texture of a beard, but such genes will only be expressed in men since hair growth is regulated by sex hormones to a large extent. Fathers and sons may have quite different beards because of alleles transmitted from their mothers as well as their fathers. In the same way there are genes transmitted through the sperm that influence breast development, but this secondary sex characteristic also is expressed in one sex under the influence of sex hormones.

Most of our traits are the outcome of multiple gene action, including the primary and secondary sex characteristics. But the simpler single-gene inheritance cases are easier to study and have been described more often.

Y-Chromosome Inheritance

There is no substantiated case of the inheritance of genes in the **Y** chromosome of humans. Earlier reports of **Y**-chromosome inheritance either have been discredited after extensive analysis or have been given different interpretations when there is enough information available. One widely cited example of **Y**-chromosome inheritance in the earlier literature dealt with a condition in humans known as *porcupine skin*, which is characterized by the formation of scales and bristles on most areas of the skin. The condition was analyzed in one family whose history could be traced back to the eighteenth century, and it was decided that the gene was not in the **Y** chromosome even though males in each generation were affected. If the gene was located in the **Y** chromosome, then we would expect that (1) *all* of the sons would resemble the father, either for the mutant or the nonmutant allele, since each son receives his **Y** chromosome from his father; and (2) the trait would *never* appear in females since they have no **Y** chromosome. These conditions were not completely fulfilled in the study that was made of the extensive family pedigree.

One of the best candidates for **Y**-chromosome inheritance is a gene that exists in an allelic form that leads to the development of hair on the outer ear (*hairy pinna*). A number of well documented family histories have been obtained, especially from southern India where this allele occurs in a higher frequency than elsewhere. One particularly complete family history did show that no female expressed the trait and that every identified male child inherited the trait from his father in a seven-generation series. But the hairiness does not appear until adulthood so that some individuals could not be identified with certainty. Since some of these identifications were crucial to the interpretation of inheritance pattern, the gaps in the pedigree have created some doubts about the conclusion for **Y**-chromosome inheritance. Additional pedigree studies may clarify the pattern, but there is no general agreement about the chromosomal location for this gene at present.

Even though we have no specific gene to assign to the **Y** chromosome, it is obvious that genes for male sex determination are present in this chromosome. A human embryo that has a **Y** chromosome begins to differentiate testes in the sixth week of its existence. Genes in the **Y** chromosome direct the development of the gonad and thus are responsible for the development of the primary sex characteristic of the male. Subsequent differentiation of male characteristics probably is regulated by genes in other chromosomes. In fact, we know of many genes in the **X** chromosome that govern a variety of traits that have no relation to sex, and we also know that autosomal genes may direct the expression of some sexual characteristics. Thus it is incorrect to assume that sexual traits are regulated only by genes in the sex chromosomes, or that genes unrelated to sex are restricted to the autosomes.

Pedigree Analysis

With humans and other species who cannot be studied using conventional and controlled breeding analysis, the size of the population sample may be too small for reliable statistical evaluation. Because we expect a 3:1 ratio in an F_2 progeny, we will not necessarily observe two classes in the exact proportion of three-fourths and one-fourth. We rarely would find 75 of one type and 25 of another in a sample of 100 F_2 individuals. We would, however, expect to find particular frequencies of all combinations *on the average* because the segregation and recombination of alleles follow random predictions based on chance happenings.

These difficulties are compounded in human pedigree studies because most of the family histories are inadequate or incomplete. Unless there are enough members of each generation, and unless we know about each member of the family, it often is difficult to make valid interpretations from the available information. Actual pedigrees do not always compare as closely as we would wish with theoretical pedigrees because of the problems of sample size and completeness.

Genetic studies based on data from family lineages or histories comprise **pedigree analysis**, in contrast with breeding analysis in which matings and progenies can be obtained according to an experimental design under standard conditions. Both kinds of studies follow the same few simple rules of inheritance that we have just discussed. Although these same rules would apply for all human traits, more effort has been directed toward the study of afflictions since major contributions have been made by medical geneticists.

Autosomal Patterns

A trait that results from the expression of a *dominant* allele of an autosomal gene ordinarily will appear in some one or more members of each generation and in approximately half of the children of a couple in which one of the parents has the trait (Fig. 2.4). When such a dominant trait is harmful to some degree or imposes some disadvantage on an individual, we find it to be very rare in human populations. When we consider the rarity of such an allele, we would further predict that most of the individuals with such a trait would be carriers, that is, having one of the dominant alleles rather than two together. In order for anyone to have two of the dominant alleles, they must have received one each from the mother's egg and the father's sperm, and this is highly unlikely for a rare allele in randomly mating human populations. For these same reasons we expect that most matings would occur between couples who totally lacked this dominant allele (hence, are recessives for the normal condition). On those rare occasions when one parent is a carrier of the trait, the second parent almost always is recessive and normal. The children of such a couple receive only the normal recessive allele from one parent, but there is a 50 percent chance *in each birth* that a child will receive the dominant allele from the afflicted parent. There is an equal chance that a child will receive the normal recessive allele from this parent, so that half of the children will be of each type *on the average*. Because only one of these dominant alleles is required for the trait to be expressed, we expect some members in each generation to display the trait since the chances are high that some of the members will have received one of these alleles.

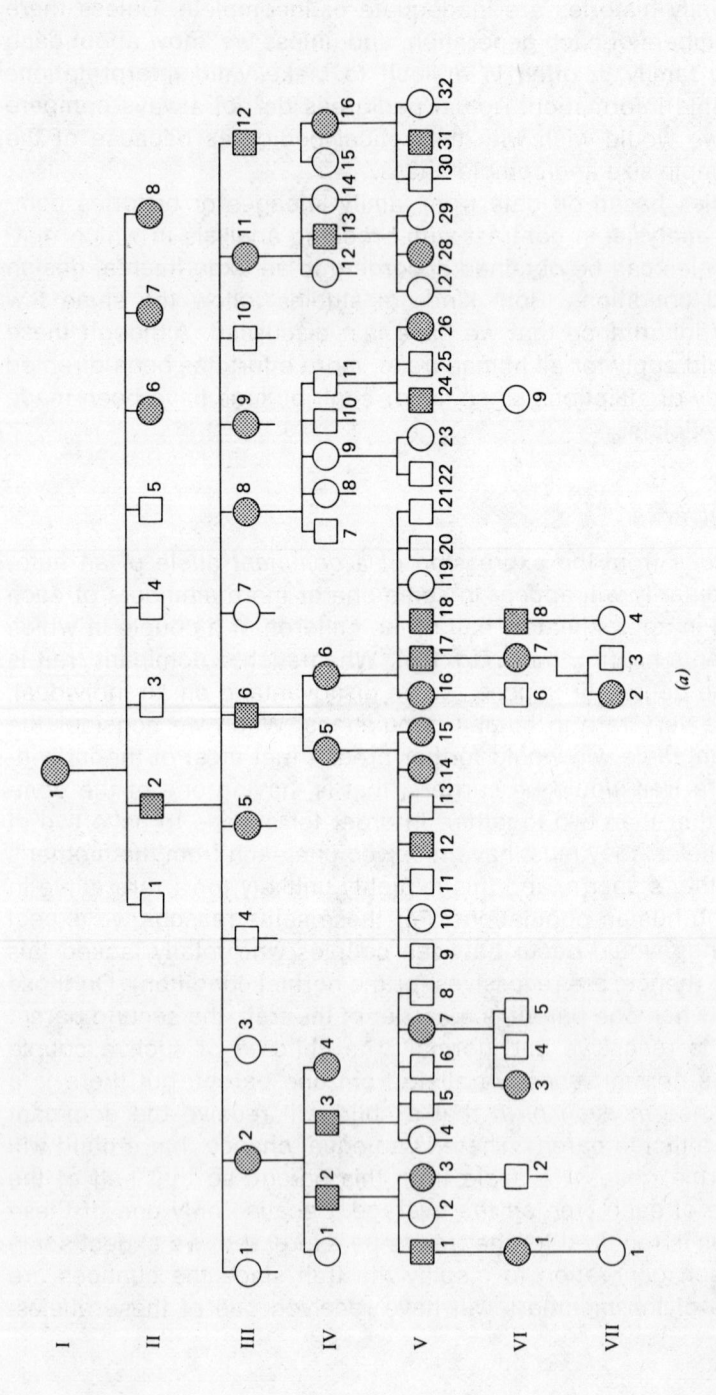

figure 2.4

Inheritance of brachydactyly (shortened fingers) in seven generations of a family. The trait does not "skip" generations, affected children are born only if at least one parent expressed the trait, and the trait appears with equal frequency in males and females; all of which indicates the inheritance of an autosomal dominant trait. Unaffected spouses are not shown in this pedigree. (From V. A. McKusick, *Human Genetics*, 2nd ed., p. 48. Reprinted with permission of Prentice-Hall, Inc., Englewood Cliffs, New Jersey. Copyright © 1969 by Prentice-Hall, Inc.)

Sex-Related Inheritance Patterns

If one of the parents happened to have two dominant alleles, then all of the children would inherit one of these and all would show the dominant trait.

Characteristics that are due to autosomal *recessive* alleles display a very different pedigree pattern. An autosomal recessive trait usually will not appear in every generation unless both parents show the trait and, most often, it will appear in children of apparently normal parents (Fig. 2.5). Again, because harmful alleles are rare in comparison with their normal alternatives, we would expect most couples to be genetically different. Matings between related individuals are relatively rare events in human populations. We can determine that both parents are carriers of the recessive allele if the trait appears in some of their children. The only way a child could inherit two recessive alleles would be for one to have come from each parent through the gametes. The theoretical prediction is that there is a 25 percent chance in each birth for the child to be of the recessive type. This prediction is based on the 3:1 ratio of dominants to recessives that we find in the (F_2) progeny of matings between carrier parents. This is the proportion that we expect to find on the average in a large population sampling, but not necessarily among the children in one family. Just because three normal children are born, it does not mean that the fourth child must be a recessive. Because there is a 25 percent

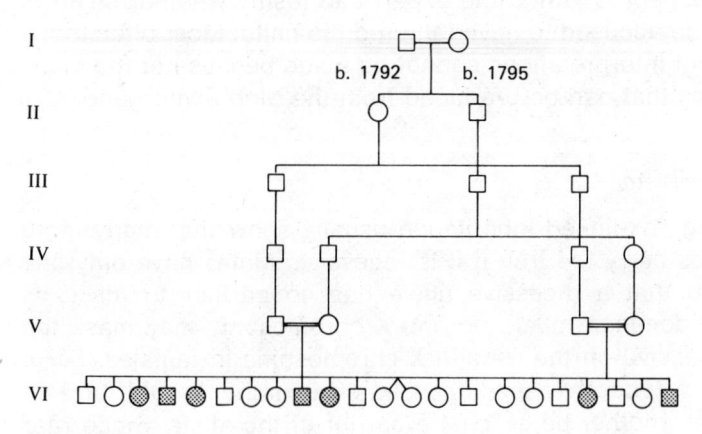

figure 2.5
The inheritance pattern of the rare affliction of hereditary microcephaly is based on an autosomal recessive allele of the gene. The immediate parents of all affected people in generation VI were descended from the same pair of great-great grandparents, and at least one of these eighteenth century relatives must have been a carrier of the microcephaly allele. (From V. A. McKusick and colleagues, *Archives of Ophthalmology*, 75 (1966), 597–600. Reprinted with permission of the publisher.)

chance of a recessive birth in each fertilization event, there may be none or one or more afflicted children in such a family, due to chance alone.

A number of common human traits show a simple single-gene inheritance pattern, and large and accurate pedigrees can be constructed to analyze the patterns. A single gene determines eye colors of brown versus blue, with brown being dominant. In analyzing such a trait, we would find that blue-eyed people have only blue-eyed children; brown-eyed couples may have only brown-eyed children, some who are brown-eyed, or none with brown eyes. These varying possibilities depend on the alleles carried by the two parents and the size of the family in realizing theoretical proportions of the two eye colors. We would be sure that no brown-eyed children would be produced by blue-eyed parents unless a rare mutational event had occurred.

These considerations form the basis for blood-type evidence which sometimes is required in paternity cases brought before the courts. There are three alleles of the gene that is responsible for the development of the major blood types: A, B, AB, and O. The O-allele is recessive to the other two, but the A and B alleles are equally dominant. A person whose blood type is AB must have received the A-allele from one parent and the B-allele from the other parent. A blood type of O can arise if the parents were of various genetic types, but not when one or both parents were of the AB blood type. By comparing the blood type of the mother and child with that of the suspected father, the expert can testify whether or not it is impossible for the accused to have fathered the child. Most often, however, such clear-cut interpretations cannot be made because of the many allele combinations that can be produced from the blood-type gene.

Sex-Linked Inheritance

Pedigrees showing sex-linked inheritance usually show that many more males than females carry the trait if it is recessive. Males have only one **X** chromosome so that a recessive allele has no partner to mask its expression, but a dominant allele on one **X** chromosome may mask the presence of a recessive on the second **X** chromosome in females. For a female to display a recessive sex-linked trait, her father would have to be afflicted and her mother be at least a carrier of the allele. Since rare alleles usually do not occur in both parents from a randomly mating population, an afflicted female is a rare event.

Sons inherit the trait from their carrier mother since their only **X** chromosome comes through the egg. For a rare allele, we usually find that afflicted males are sons of seemingly normal mothers, but these women carry a recessive allele in addition to the dominant one which leads to their own normal development and appearance. Females may receive a recessive mutant allele either from their mother or father since each parent contributes one of the **X** chromosome pair through the gametes.

One of the more interesting pedigrees of "classical" hemophilia is that of Queen Victoria of Great Britain and five generations of her descendants scattered among the royal houses of Europe (Fig. 2.6). Of her five daughters, two definitely were carriers of the trait, two probably had only the normal dominant alleles since there was no evidence of the trait in their descendants, and Princess Louise had no children so we cannot be sure of her genetic constitution. One of the four sons (Leopold) born to Queen Victoria and Prince Albert was a hemophiliac, and the other three sons showed normal blood-clotting ability. The British royal family escaped the disease since King Edward VII was one of the three non-hemophiliac sons, thus all of his descendants were normal. In this particular pedigree as in many others for recessive sex-linked traits, only males display the affliction. None of the females in the families were born to carrier mothers married to hemophiliac men, so that no case occurred in which a recessive allele from both parents produced a recessive hemophiliac daughter.

The allele for hemophilia probably arose as a mutation in Queen Victoria since there is no evidence of its occurrence in her parents, grandparents, or other ancestors. The protocol for royalty marrying royalty, and for relatively large families, contributed to the downfall of some of the European royal houses to which the hemophilia allele was transmitted. The best known instance of the influence of this allele on European history involves the Russian royal family, none of whom remained alive after the 1917 revolution. The Tsarina was a granddaughter of Queen Victoria, and she inherited the hemophilia allele from her mother, who received the recessive allele from Queen Victoria. The Tsarina was a carrier, like her mother and grandmother before her, and none of these three women displayed the "bleeder's disease" since each had a second **X** chromosome with a normal dominant allele that masked the recessive. The Tsar and Tsarina had five children, four daughters and their son, the Tsarevitch Alexis, a hemophiliac. During his childhood, Alexis suffered serious crises of bleeding, and his parents were constantly on the lookout for any remedy to help their son. One of their actions led to their dependence on the scoundrel Rasputin since this man appeared to have some power to limit or stop the bleeding episodes suffered by their son. Rasputin's power extended to various areas of political activities, and he came to exercise considerable influence in the country. According to some versions of this history, the evil influence of Rasputin may have been one of the precipitating factors that brought on the Russian revolution in 1917. He would never have become an influential person if it had

44 GENETICS AND EVOLUTION

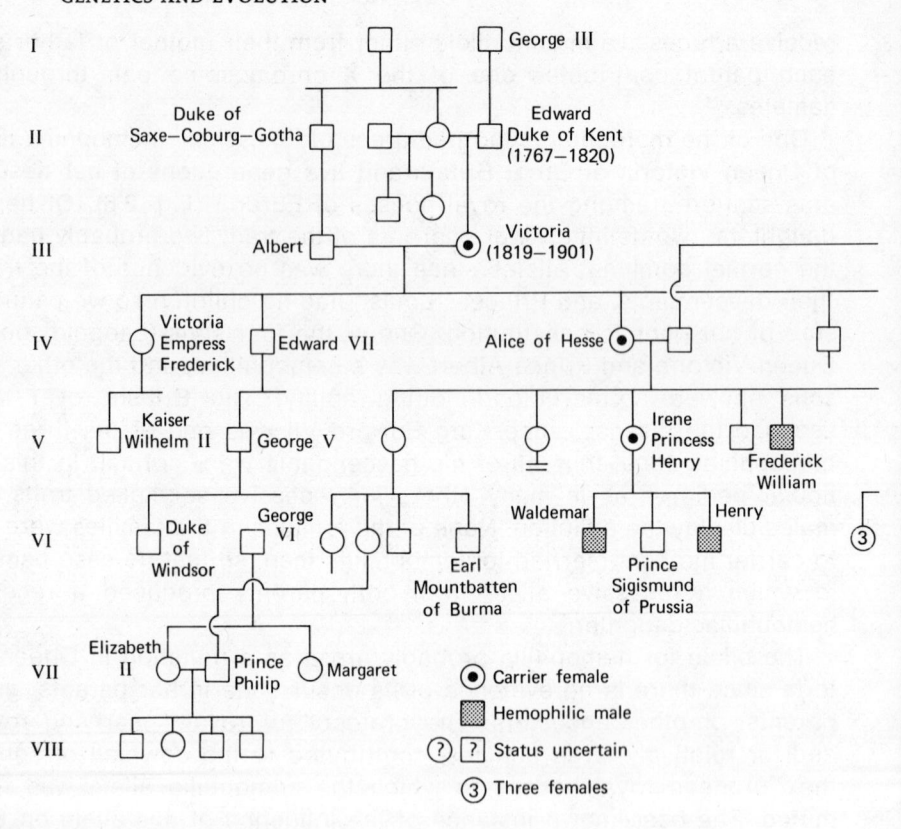

figure 2.6
The inheritance of "classical" hemophilia through Queen Victoria of Great Britain. There are no female hemophiliacs, since no couples occurred in which the mother was a carrier and the father was a hemophiliac. This pedigree clearly demonstrates sex-linked inheritance of a recessive trait. (From V. A. McKusick, *Human Genetics*, 2nd ed., p. 56. Reprinted with permission of Prentice-Hall, Inc., Englewood Cliffs, New Jersey. Copyright © 1969 by Prentice-Hall, Inc.)

not been for his treatments of the Tsarevitch, and perhaps history would have been somewhat different under other circumstances.

Hemophilia A, the "classical" form of the disease, is due to the action of one gene among the nine or ten that are known and that are believed to act in the production of the thirteen or more substances required for blood to clot in the normal time of about five minutes. It may take two hours for clotting to take place in a person suffering with hemophilia A because of the lack of the plasma protein under the control of this gene. On the basis of pedigree studies, about seventy other sex-linked genes

Sex-Related Inheritance Patterns

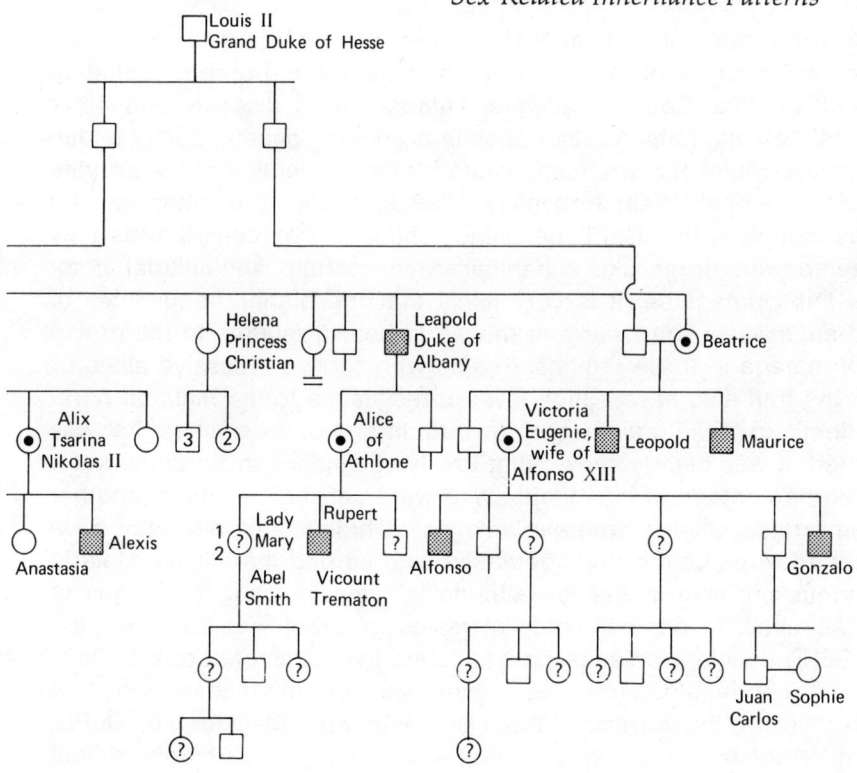

[Pedigree chart of the descendants of Louis II, Grand Duke of Hesse, showing the inheritance of hemophilia through European royalty, including Helena Princess Christian, Leopold Duke of Albany, Beatrice, Alix Tsarina (wife of Nikolas II), Alice of Athlone, Victoria Eugenie (wife of Alfonso XIII), Leopold, Maurice, Anastasia, Alexis, Lady Mary, Rupert, Abel Smith, Vicount Trematon, Alfonso, Gonzalo, Juan Carlos, and Sophie.]

have been identified in humans. Red-green colorblindness is another of the better known sex-linked conditions, but there are different degrees of colorblindness, and more than one sex-linked gene is responsible for the development of the condition. About 75 percent of colorblind people are insensitive to green and 25 percent are insensitive to red, in varying degrees, depending on the genes that are present. Total colorblindness is extremely rare. In the United States, about 8 percent of the males have one or another of the types of red-green colorblindness, but only 0.4 percent of the female population shows this condition.

There are a number of hemoglobin diseases that occur in fairly high frequencies in parts of Africa, Asia, and southern Europe; including sickle-cell anemia, Cooley's anemia, hemoglobin C disease, and other autosomal genetic traits. Another anemia-producing genetic defect is due to sex-linked alleles that are responsible for the production of the enzyme glucose-6-phosphate dehydrogenase (G6PD). Unlike the other genetic anemias, people with a G6PD deficiency show no symptoms unless they are treated with drugs like sulfanilamide or certain antimalarial compounds like primaquine. It is very likely that the higher frequencies of genetic anemias in some parts of the world bear a relation to the prevalence of malaria in these regions. People who carry a recessive allele or display the trait also have a high level of resistance to the malarial parasite endemic to these regions of the world. In one of the studies that was conducted, it was clearly shown that low-lying villages in Sardinia had a very high percentage of inhabitants who were carriers of one or another of these anemia alleles, whereas villages at higher altitudes may have had only a few percent of the population who carried the alleles. Malaria is a serious problem in the low-altitude villages, but not in villages at higher altitudes. In another study of males in Israel who had the sex-linked G6PD deficiency disease, it was found that Jews of northern European extraction included only two percent showing the disease, whereas sixty percent of the Kurdish Jews from Asia were deficient in G6PD. These differences in frequency of alleles in different populations reflect the relative advantages and disadvantages of a genetic trait in a particular environmental situation. The mutation probably has occurred in all populations at various times, but the alleles increased in frequency only in those populations where they conferred some advantage on the individual. African blacks who were brought to this country in slave ships two to three centuries ago almost certainly showed a higher proportion of carriers of the sickle-cell trait than we now find in American black populations. The probable original proportion of carriers was 22 percent, according to information from present-day African populations and from other sources, but American blacks today include only nine or ten percent carriers of the trait. The disadvantage of the allele in America contributed to its reduction in frequency over the years. Continued advantage in malarial regions of Africa contributes to the maintenance of a higher carrier frequency there.

Although most of the known sex-linked traits are recessive, there are some rare dominant traits due to genes on the **X** chromosome. One of the better known of these sex-linked dominants is the allele which is responsible for the development of skeletal deformities due to vitamin D-resistant rickets. In the case of an **X**-linked dominant, we find that twice as many females display the trait as compared with males. Females have two **X** chromosomes, and the presence of even one dominant allele will lead to the development of the trait. Since there are two **X** chromosomes, the female has twice the chance of inheriting an **X** chromosome with the dominant allele and thus is twice as likely to be afflicted. She may inherit the allele either from her mother or her father, whereas a male inherits his only **X** chromosome exclusively from his mother.

Suggested Readings

Ayer, A. J., Chance. *Scientific American, 213* (October 1965), 44.

Hunter, R., and I. Macalpine, Porphyria and King George III. *Scientific American, 221* (July 1969), 38.

Kac, M., Probability. *Scientific American, 211* (September 1964), 92.

Lerner, I. M., *Heredity, Evolution, and Society.* San Francisco: Freeman, 1968.

McKusick, V. A., The royal hemophilia. *Scientific American, 213* (August 1965), 88.

Winchester, A. M., *Human Genetics.* Columbus, Ohio: Charles E. Merrell Publ., 1971.

chapter three

Human Medical Genetics

There is a great emphasis today on medical genetic problems in which the condition is due to some change in a chromosome rather than a particular gene in a chromosome. These studies were not possible before 1956 because we had no useful methods for preparing and examining human chromosomes under the microscope. In fact, it was generally believed that human cells contained 48 chromosomes, although the preparations were of such poor quality that it was difficult to be sure of the interpretations that were made. We knew little about any of the human chromosomes, including the **X** and **Y**. The development of new methods in 1956 opened the way to successful and relatively simple approaches to the study of human chromosomes in ordinary body cells, such as those in the blood. Studies since 1956 have provided very clear evidence showing that there are 46 chromosomes in human body cells, with 44 of these being autosomes and the other two either **XX** in females or **XY** in males.

The Human Chromosomes

Cells for study with the microscope can be obtained from various sources such as skin, blood, or specimens taken for biopsies in the hospital. Relatively quick chromosome preparations can be made from small quantities of blood that is removed by a trained person. The white blood cells are separated from the rest of the sample and are treated in special ways to make the chromosomes visible and distinguishable in considerable detail. These cells ordinarily are not mitotic so that the chromosomes in the nucleus look tangled and impossible to identify. The special treatment causes many of these cells to begin to divide by mitosis, which leads to a shortening of the chromosomes so that they are more distinct. A chemical is added to stop the mitosis in the metaphase stage (see Fig. 1.8), when the chromosomes are at their most contracted size and are well separated from one another. An additional step leads to swelling of the cells, and this causes the chromosomes to spread apart even more so that maximum visibility of each chromosome becomes possible. These preparations then are scanned under the microscope, and suitable chromosome sets are photographed with the camera that is attached to the microscope. After the photograph is enlarged, the individual chromosomes usually are cut out and pasted on a sheet in a particular order, starting with the largest and ending with the smallest chromosomes in the set (Fig. 3.1). Since each body cell contains one set of 23 chromosomes derived from the mother and one set of 23 from the father, there actually are 23 chromosome *pairs* that make up the 46 that can be counted. The chromosomes, therefore, are paired and arranged according to their size and other features of their shape and form. The **X** and **Y** chromosomes are identified by their letter designations, but the autosomes are numbered from 1 to 22 and referred to by their number; for example, as chromosome-5 or chromosome-21, and so forth. This is the most convenient way to discuss the human chromosomes and various clinical conditions in which chromosome modifications are implicated.

The Sex Chromosomes

The **X** chromosome is of intermediate size and is usually difficult to distinguish from autosomes of similar size and shape. Similarly, the **Y** chromosome resembles the smallest autosomes, chromosome-21 and chromosome-22. There is evidence showing that the size of the **Y** chromosome may vary among males in a population but that the chromosome generally is similar among males from the same family. The significance of this observation is unknown. With new methods that recently have been developed to stain the chromosomes differentially, it is much simpler to identify each chromosome of the complement with greater accuracy when photographed through the microscope.

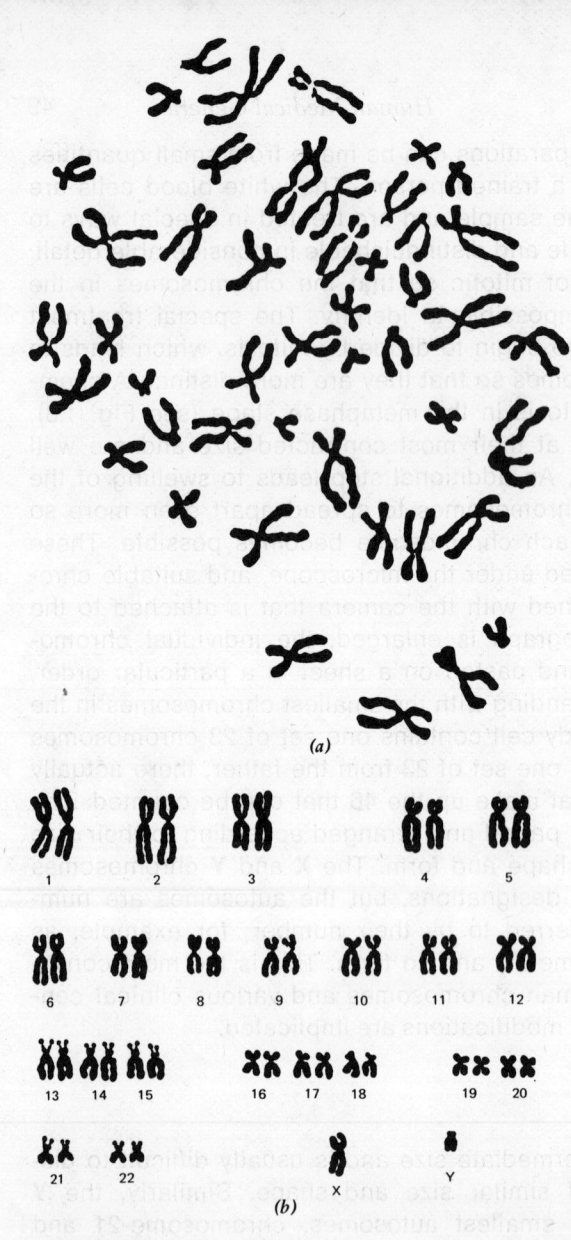

figure 3.1
Human chromosomes from body cells that were prepared to arrest division and then to obtain maximum spreading out of the 46 chromosomes. (a) Appearance of a chromosome complement from a specially prepared cell. (b) The conventional arrangement of the 23 pairs of chromosomes taken from a photograph such as shown in (a), in order of descending sizes. The X chromosome is intermediate in size, whereas the Y chromosome is among the smallest present.

Sex Determination

There were various proposals to explain the sex chromosome basis of sex determination in humans and other mammals, but one that was very widely accepted in earlier years was based on studies of insects such as the fruit fly and grasshopper. In those insects, the **Y** chromosome seems to be unnecessary for sex determination, and many insects lack a **Y** chromosome altogether. The sex determining mechanism apparently depends on the number of **X** chromosomes in proportion to the two sets of autosomes that also are present in insect body cells. Insects that had one **X** chromosome developed as males, and insects with two **X** chromosomes developed as females. Because there also are **XX**-female—**XY**-male patterns in some of these insect groups, it was assumed that the mammalian **XX**-female—**XY**-male sex determining system also was based on this pattern of one versus two **X** chromosomes, and that the **Y** chromosome was irrelevant to sex determination.

After 1956, when human chromosomes and those from other mammals could be studied accurately, a great deal of evidence was accumulated that showed that both mammalian **X** and **Y** chromosomes were necessary for sex determination. The system is quite unlike the **XX**—**XY** mechanism in insects, even though the same kinds of sex chromosomes may be present in all of these groups. When there is only one **X** chromosome and no **Y**, the insect develops as a male. When this **XO** (pronounced "X-Oh") condition occurs in humans, then the individuals who develop are female. This line of evidence by itself was enough to throw suspicion on the earlier notions about the similarity between insect and mammal sex chromosomes and sex determination. But additional evidence soon was obtained that definitely showed that the **Y** chromosome was the essential male determinant no matter how many **X** chromosomes also might be present in a person.

There is no known incidence of development in humans when the **X** chromosomes are missing and no known case where only a **Y** chromosome occurs and the **X** is absent. The embryo apparently can tolerate the loss of one **X** chromosome if a second **X** is retained, but it cannot survive to any stage of development without at least one **X**. Embryos obviously may thrive in the absence of the **Y** chromosome, since all females lack this chromosome, so that there cannot be any genes that are vital to life on the **Y**. If the embryo has more than two **X** chromosomes and lacks a **Y**, then female development takes place. If there is a **Y** chromosome, then male development takes place whether there is one **X**, as in normal males, or more than one **X** chromosome, as in a high frequency of males in the human population. Since individuals who were **XY**, **XXY**, **XXXY**, and **XXXXY** all developed as males, it was obvious that the **Y** chromosome was strongly male-determining in humans (and other mammals). Although each person who has a **Y** chromosome is male, the presence of extra **X** chromosomes leads to sterility and to the differentiation of some feminized traits in such men (Table 3.1).

Barr Bodies

Although it was difficult to identify the human sex chromosomes until improved methods were developed, there was one line of evidence that showed differences in the nucleus of cells from males and females. When cells were taken from some part of the body, usually by scraping the lining of the cheek inside the mouth, they could be examined with the microscope after suitable preparation and staining with dyes. In these nondividing nuclei from females, there was a dense clump of heavily stained chromosome material called a **Barr body**, in recognition of its discoverer, Murray Barr. Cells from males lacked the Barr body (Fig. 3.2). Patients who had some form of sexual anomaly often showed a different Barr body pattern; for example, some males possessed a Barr body and some females lacked a Barr body in the nuclei of their cells. After 1956, when sufficient information had been collected from direct observations of the chromosomes, the pattern of Barr body distribution clearly was shown to coincide with the number of **X** chromosomes in the cell. In every case, there is one less Barr body than number of **X** chromosomes. Normal females with two **X** chromosomes have one Barr body per

table 3.1

Sex Chromosome Anomalies and Sex Determination in Humans

Individual Designation	Chromosome Constitution[a]	Total Number of Chromosomes	Sex	Fertility
Normal male	AA XY	46	Male	+
Normal female	AA XX	46	Female	+
Turner syndrome	AA X	45	Female	−
Triplo-X	AA XXX	47	Female	±
Textra-X	AA XXXX	48	Female	?
Penta-X	AA XXXXX	49	Female	?
Klinefelter syndrome	AA XXY	47	Male	−
Klinefelter syndrome	AA XXXY	48	Male	−
Klinefelter syndrome	AA XXXXY	49	Male	−
Klinefelter syndrome	AA XXYY	48	Male	?
Klinefelter syndrome	AA XXXYY	49	Male	?
XYY-male	AA XYY	47	Male	+

[a] AA refers to two sets of autosomal chromosomes; X and Y refer to the actual number of each sex chromosome in a body cell.

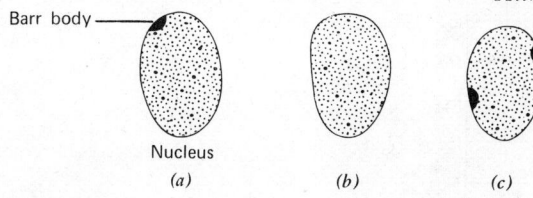

figure 3.2
The nondividing nucleus of some kinds of body cells often shows a Barr body, which is a highly condensed X chromosome. The number of Barr bodies in the nucleus always is *one less* than the total number of X chromosomes in the particular complement that is present. (*a*) One Barr body would be found in normal XX females and in Klinefelter males who were XXY or XXYY; (*b*) No Barr bodies occur in nuclei of normal XY males, XO females with Turner's syndrome, or XYY males; and (*c*) Two Barr bodies would occur typically in individuals who possessed three X chromosomes, such as XXX females or XXXY males with Klinefelter's syndrome.

nucleus; normal males with only one **X** chromosome lack the Barr body. Patients with sex chromosome variations provided further evidence to support this relationship between numbers of Barr bodies and **X** chromosomes in the nucleus. The **XO** female had no Barr body; **XXY** males showed one Barr body; **XXX** females showed two Barr bodies per nucleus; **XXXY** males had two Barr bodies; and so forth. The Barr body apparently corresponds to a highly condensed **X** chromosome, and each **X** chromosome in excess of one will become condensed and appear as a Barr body when it is viewed through the microscope.

In many population studies and clinical cases, it is desirable to know the sex chromosome constitution. This information can be obtained very quickly and inexpensively by examining cells from the cheek lining and counting the number of Barr bodies. The alternative of preparing blood cells so that the individual chromosomes can be photographed, identified, and counted is time consuming and relatively expensive. Unless such specific information is required for treatment or for research purposes, the simplest procedure is to examine cells for Barr bodies and determine the number of **X** chromosomes in the individual. If the individual is a male, then at least one **Y** chromosome must be present; if the person is female, then a **Y** chromosome must be absent. Barr body counts together with the sex identification of the individuals can provide useful information concerning the frequency of chromosome anomalies in the population as a whole or in localized groups of people.

Sex Chromosome Anomalies

About one in 400 to 600 live male birth results in an individual with the clinical symptoms of **Klinefelter's syndrome** (Fig. 3.3). These males are

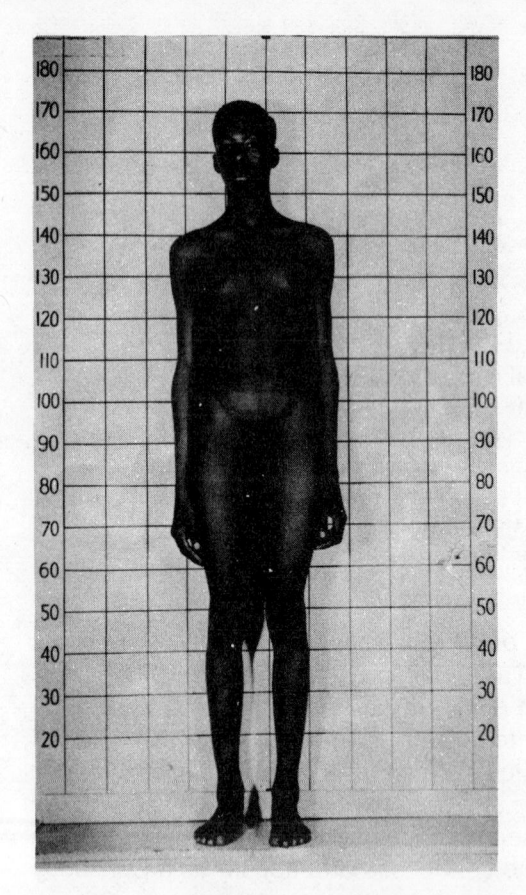

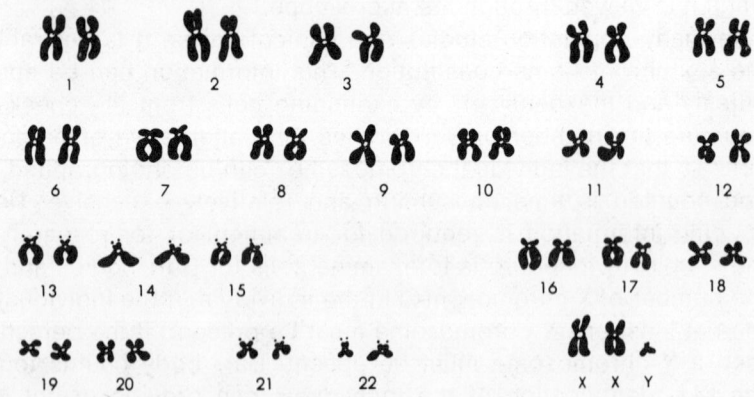

figure 3.3
A male showing the characteristics of Klinefelter's syndrome: small testes, sparse body hair growth, femalelike breast development, and long limbs. There are 47 chromosomes in his body cells, since XXY is present along with 44 autosomes. (Reprinted with permission from the *Journal of Chronic Diseases*, July, 1960.)

sterile, and they have such feminized features as enlarged breasts and sparce hair growth. The external genitalia are male, but the penis and testes are small. Arms and legs are disproportionately long in such males but, more importantly, most of these individuals are mentally retarded to some degree. Most cases of Klinefelter's syndrome are **XXY** and have 47 chromosomes, but there may be as many as four **X** chromosomes, so that the number varies depending on the particular individual. The number of extra **X** chromosomes can be determined quickly by examining cells for Barr bodies. The normal male has no Barr body, so that the number that is found can be translated directly into the number of extra **X** chromosomes present. If a more detailed analysis is required, then the chromosomes must be examined by the usual procedure that we discussed earlier in this chapter. The **Y** chromosome acts early in embryo development so that testis development is initiated and a male sex is determined. The clinical symptoms that are displayed later probably are due to the action of female sex hormones in conjunction perhaps with lowered effectiveness of male sex hormone in these cases.

Recently, there has been some study of unusually tall and aggressive males whose chromosome analysis reveals an extra **Y** chromosome in the total count of 47 chromosomes. Some of these men have criminal records, which first brought them to attention. From subsequent studies of different groups of men in prisons, hospitals, and the general population, it is clear that **XYY** males may have normal intelligence as measured by IQ tests and show normal social adjustments, but many are somewhat retarded mentally and show antisocial tendencies that we label as criminal. The correlation between the **XYY** chromosome constitution and the higher than average frequency of these men in penal institutions has led to the question of cause and effect. At present, we do not know whether their antisocial actions are due to inborn problems related to the chromosomes or to the psychological trauma that may result when such boys interact with their peers. They grow very tall and may not be as quick and bright as other boys, all of which could cause them to withdraw from the usual social affiliations and lead to overt hostilities. These questions must be evaluated especially in legal cases involving **XYY** male defendants, but no definitive statements can be made at present. The exact proportion of **XYY** males in the population is unknown, but it has been estimated that 1 in 2000 male births may be **XYY**. For some (and possibly all) **XYY** males there is no sterility, and **XYY** males are known to have fathered children. There is a great difference between **XXY** and **XYY** males even though each has 47 chromosomes because an extra sex chromosome is present. The chromosome unbalance leads to more profound effects when the extra chromosome in the male is an **X**.

Sex anomalies among females are of two principal types: the more frequent individuals (1 in 1400 births) with more than two **X** chromosomes and those who exhibit the clinical symptoms of **Turner's syndrome** (1 in 3500 live births). We know much more about Turner's syndrome because many of these females require or receive medical care. Females with this syndrome are **XO** and, therefore, they have only 45 chromosomes and show no Barr body in the cell nucleus (Fig. 3.4). The external genitalia are female, but the internal reproductive tract is undeveloped and there is little breast enlargement in the adults. In addition to a number of physical symptoms, about which there is little controversy, there have been conflicting reports about the level of intelligence of **XO** females. Some reports have cited mental retardation in **XO** females, while other reports have stated that the particular deficiencies shown in aptitude tests do not necessarily imply retardation. For example, some **XO** females have a relatively poor aptitude for arithmetic, but this does not necessarily influence their total mental capacity to function. Unlike the multiple-**X** females and Klinefelter males who have a high chance for live birth, **XO** female embryos have a reduced chance. Approximately five to six percent of spontaneous abortions have been found to involve a fetus with the **XO** chromosome constitution of Turner's syndrome. It is more dangerous for the fetus to have fewer than the normal number of chromosomes than to have more than 46 chromosomes.

Autosomal Chromosome Anomalies

Changes in Chromosome Number

The first abnormality involving an autosomal chromosome that was described in humans was the presence of an extra chromosome-21 in males and females with the symptoms of **Down's syndrome** (mongolism). These individuals have 47 chromosomes instead of 46, a characteristic set of physical features (Fig. 3.5), and a degree of mental retardation that may be quite severe in some cases. Approximately 1 in every 500 or 600 live births results in a mongoloid child, but there also is a high rate of spontaneous abortions when the fetus is of this chromosome constitution, especially among older mothers. Women with Down's syndrome have given birth to children, and the studies show that the child also was mongoloid in approximately half of the cases. This is expected because the oocytes that produce the egg cells after meiosis have 47 chromosomes; when the chromosomes in these oocytes are reduced in number and segregated into the gamete cells, half of the time the gamete might receive the usual set of 23 chromosomes and half of the time the chances are that the extra chromosome-21 might also be included in the egg. The distribution of the extra chromosome occurs at random, but it is transmitted to the offspring in 50 percent of the births on the average.

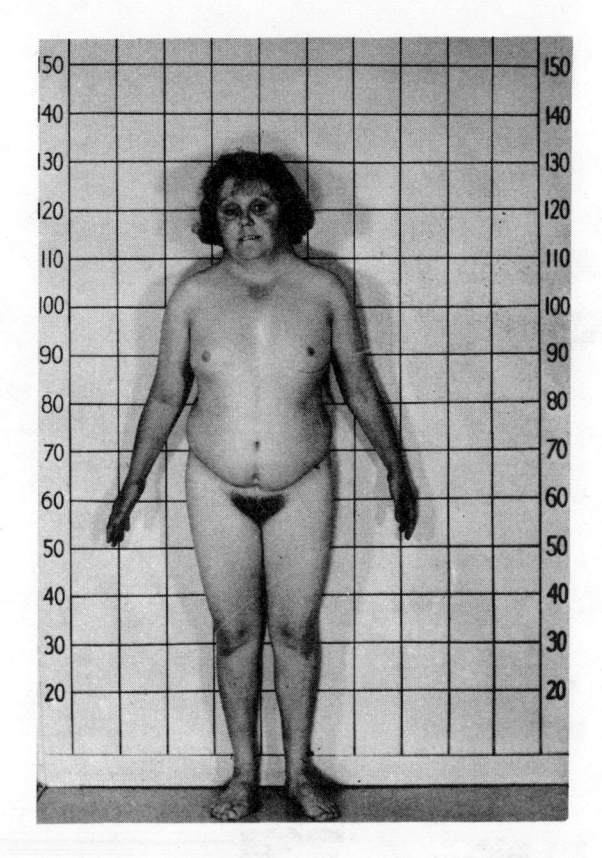

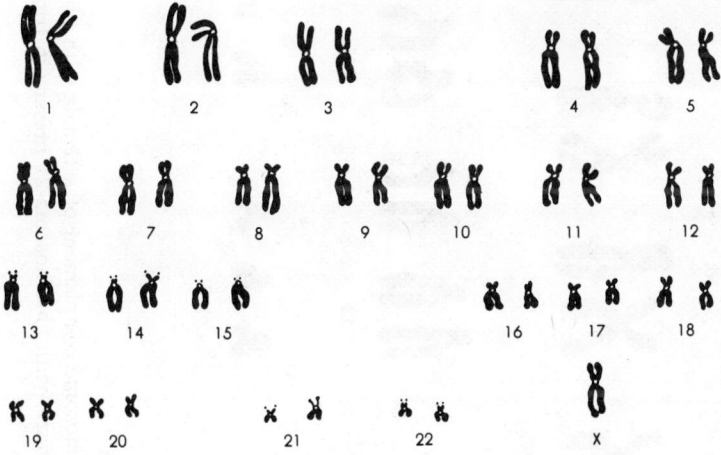

figure 3.4
A female with the characteristic features of Turner's syndrome. Chromosome analysis revealed 45 chromosomes, with one X chromosome missing from the complement. (Reprinted with permission from the *Journal of Chronic Diseases*, July, 1960.)

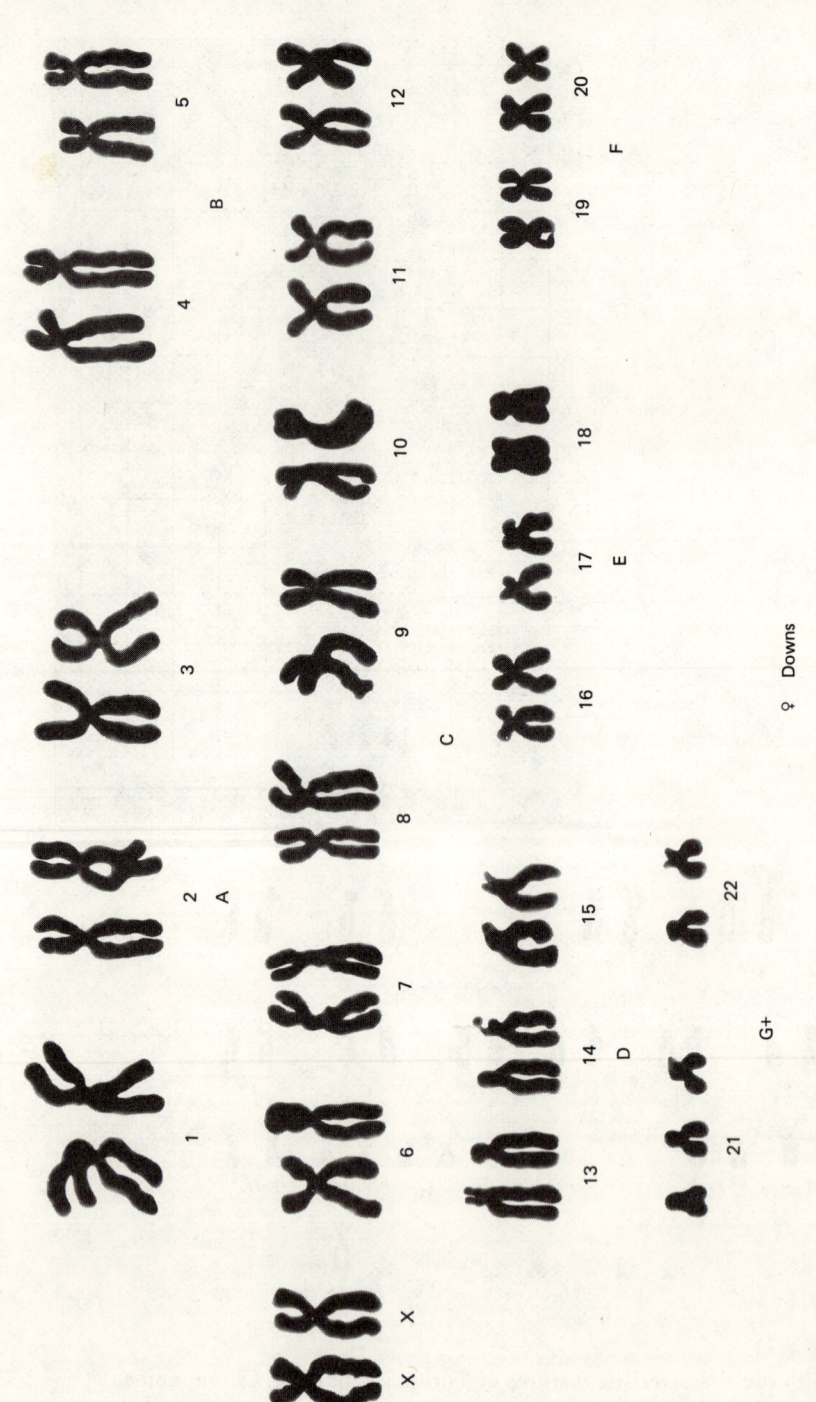

figure 3.5
The chromosome complement of a female with the symptoms of Down's syndrome, or mongolism. There are 47 chromosomes, with chromosome-21 present in triplicate instead of the usual pair. (Photograph courtesy of L. J. Sciorra.)

Until recently, it was believed that the principal environmental influence leading to the birth of a mongoloid child was the age of the mother at the time the infant was conceived and born. There is clear evidence showing that the incidence of mongoloid births is much higher for women between the ages of 40 and 50. There is a steady increase in the frequency of such births for mothers of increasing age (Fig. 3.6), especially between the ages of 40 and 45. Some information has been provided recently that raises the possibility that virus infection may increase the chance for birth of a child with Down's syndrome. Studies of some Australian populations have shown that such births may occur at particular times and spread to other geographical areas, very much resembling the pattern of occurrence and spread of virus infections. The full nature of these relationships is unknown, and we have relatively little information in general about the interactions of genetic and environmental factors that influence the observed patterns of Down's syndrome births.

Two other conditions have been reported that involve an extra autosome, one involving chromosome-13 and the other showing an extra chromosome-18. The infants do not survive for very long in either of these cases, although they are not particularly rare births. About one in 4,500 live births shows an extra chromosome-18, and about one in 14,500 has an extra chromosome-13 among the five million babies born each year in the United States. Each of these conditions also contributes to the rate of spontaneous abortions, about three percent showing the extra chromosome-13.

The only viable condition in which one of the chromosomes is missing is the **XO** female with 45 chromosomes. There is no known case in which an aborted fetus showed a missing autosomal chromosome. Apparently, the level of genic unbalance is so severe when one of the autosomes is missing that abortion either occurs very early and is not detected, or the gamete that carries such an unbalanced chromosome complement is inviable and never achieves the stage of fertilization.

Changes in Chromosome Structure

Some patients with Down's syndrome have 46 chromosomes instead of the expected 47. Careful study of the chromosomes shows that there are two of the chromosome-21 clearly visible, but the third chromosome-21 has become attached to one of the other autosomes, usually chromosome-15 (Fig. 3.7). These patients actually have three of the chromosome-21, plus all the rest of the usual chromosome complement, and they show the same symptoms as the more common cases in which all three of the chromosome-21 are separate and identifiable in cells with 47 chromosomes. One of the parents, usually the mother, has this same unusual chromosome-21 attached to chromosome-15 and thus shows only

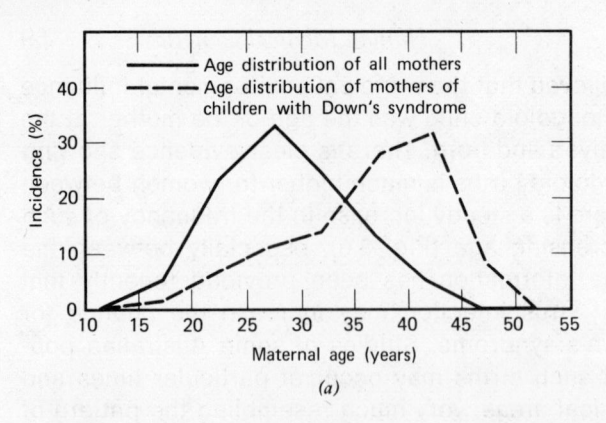

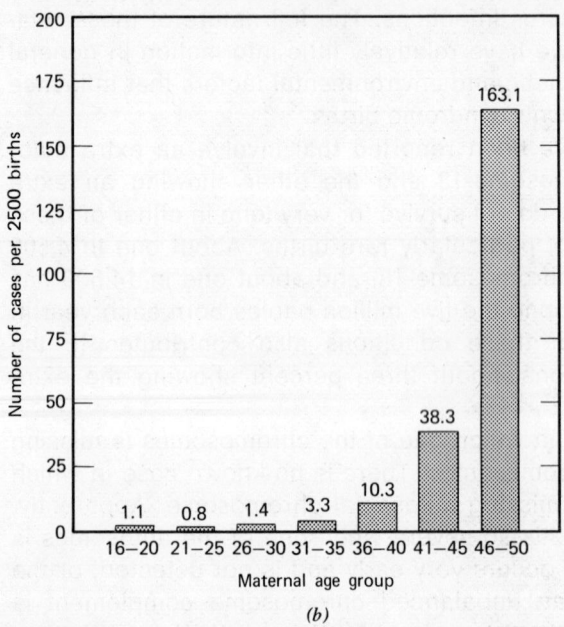

figure 3.6
There is an increasing frequency of children with Down's syndrome born to mothers in the older childbearing years. (a) Although affected children are born to mothers of various ages, the greatest percentage occurs for mothers who are over 30 years of age. (b) The absolute frequency of children with Down's syndrome increases with the increasing age of the mother and is dramatically higher for women in the 46–50 age group.

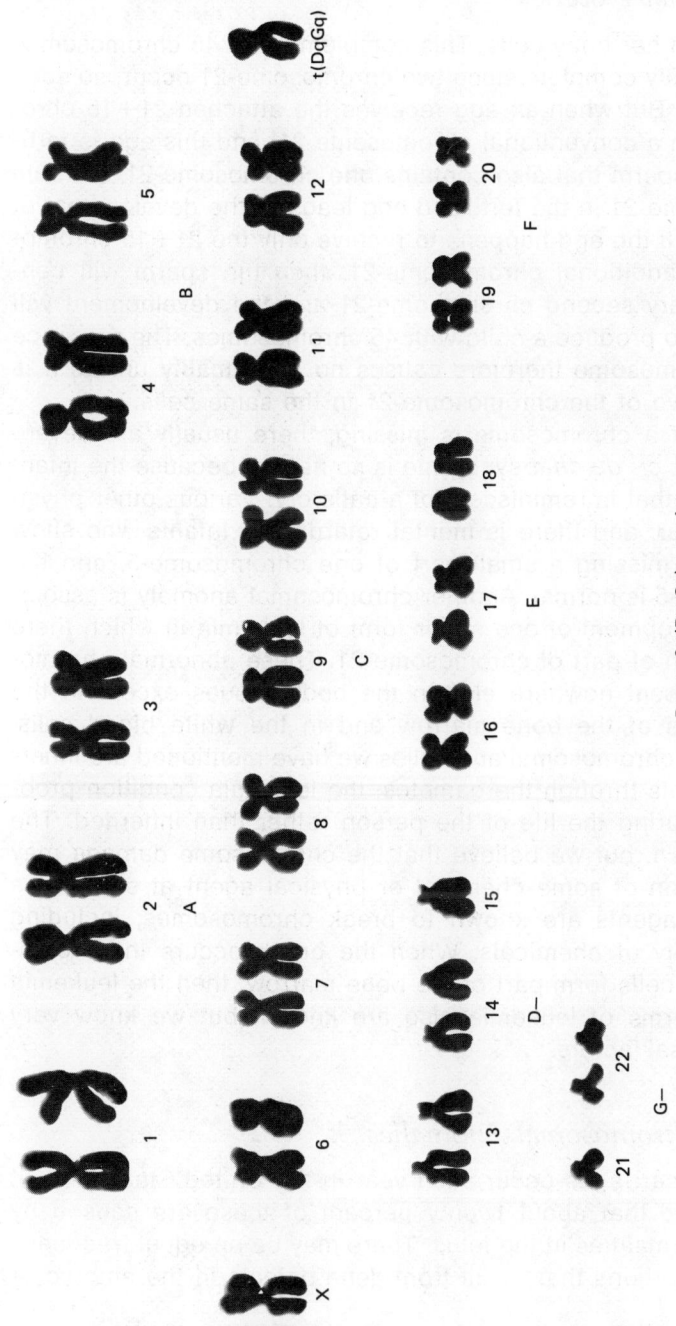

figure 3.7

In some cases, the mother or other relatives of a patient with Down's syndrome may have only 45 chromosomes and be normal in appearance. An examination of the chromosomes reveals that one of the 45 chromosomes actually is composed of two chromosomes that form a single structure. These usually are a combination of chromosome-15 plus chromosome-21. Note that there is only one recognizable chromosome-15 and one chromosome-21, in addition to the unusual 15+21 chromosome structure. These individuals possess a full set of genes. (Photograph courtesy of L. J. Sciorra.)

45 chromosomes in her body cells. This complement of 45 chromosomes actually is genetically complete since two chromosome-21 occur, so such people are normal. But when an egg receives the attached 21+15 chromosome along with a conventional chromosome-21, and this egg is fertilized by a normal sperm that also contains one chromosome-21, the sum of three chromosome-21 in the fertilized egg leads to the development of a mongoloid child. If the egg happens to receive only the 21+15 chromosome and not an additional chromosome-21, then the sperm will contribute the necessary second chromosome-21 and the development will proceed normally to produce a child with 45 chromosomes. The presence of the 15+21 chromosome therefore causes no abnormality unless it is accompanied by two of the chromosome-21 in the same cells.

When a piece of a chromosome is missing, there usually are severe consequences. The *cri du chat* syndrome is so named because the infant has a plaintive cry that is reminiscent of a cat's cry. Various other physical symptoms occur, and there is mental retardation. Infants who show this syndrome are missing a small part of one chromosome-5, and the other chromosome-5 is normal. Another chromosomal anomaly is associated with the development of one major form of leukemia in which there has been a deletion of part of chromosome-21. These abnormal chromosome sets are present nowhere else in the body tissues except in the blood-forming cells of the bone marrow and in the white blood cells. Although the other chromosomal anomalies we have mentioned are inherited from the parents through the gametes, the leukemia condition probably is acquired during the life of the person rather than inherited. The causes are unknown, but we believe that the chromosome damage may be due to the action of some chemical or physical agent at some time during life. Many agents are known to break chromosomes, including x-rays and a variety of chemicals. When the break occurs in chromosome-21 and such cells form part of the bone marrow, then the leukemia develops. Other forms of leukemia also are known, but we know very little about the causal factors.

Significance of Chromosomal Anomalies

About 500,000 miscarriages occur each year in the United States, and it has been estimated that about twenty percent of these are caused by chromosome abnormalities in the fetus. There may be an equal frequency of spontaneous abortions that result from gene defects in the embryo or fetus.

At least 20,000 of the babies that are born each year in the United States have some physical or mental defect due to chromosomal abnormalities. As mentioned earlier, the birth rate of infants with Down's syndrome is about one in 500 or 600, and for Klinefelter males it is one in 400 to 600 live male births. For a birth rate of five million babies per year, it means that about 10,000 newborns will be mongoloid, about 5,000 males each year are born with Klinefelter's symptoms, and the remainder will bear one or another of the known afflictions caused by chromosomal abnormalities. The emotional, societal, and financial burdens that are imposed by these conditions are substantial. Some of these genic and chromosomal defects can be predicted to occur for certain couples and families, or a probability of risk of the occurrence can be estimated. For some afflictions, it is possible to determine the presence or absence of a particular condition in the young fetus, with the most suitable times being between twelve and sixteen weeks after conception. It is possible to make an earlier prenatal diagnosis, or to determine the condition in the fetus as late as the fifth month of pregnancy. We will now discuss some of the prospects for prenatal genetic diagnosis and genetic counseling, which have become more generally known to the public in recent years.

Medical Genetics

By collecting information from prospective parents about the genetic history of the family, it is possible to construct a pedigree that shows the inheritance pattern for some condition. But we know the inheritance patterns for numerous genetic conditions, so that this basic information can be used to provide the family with an estimate of risk in having a child with some particular genetic characteristic. If we know the pattern of inheritance and we learn the probable genes that are present in the prospective parents, then we may be able to predict the kinds of gametes that are produced and the chances of appearance of a genetic condition after conception.

Determination of Risk

We know that Tay-Sachs disease (infantile amaurotic idiocy) develops in infants who have two recessive alleles of the gene that is involved. Such children only can be produced when both parents carry the recessive allele and if the egg and the sperm that fuse happen to carry this allele at the same time. The condition does not develop in the carriers of the allele, and they may be quite unaware of the presence of the Tay-Sachs recessive allele since the normal dominant allele masks the expression of the disease. The disease is lethal, causing death of the child before the age of four. The birth of infants with Tay-Sachs disease can occur only when both parents are carriers. If only one parent carries the recessive allele, then the other parent provides a masking dominant allele in every gamete, and the resulting fertilized egg will develop normally into a healthy person. Therefore, the couples most concerned about the risk of a pregnancy leading to a Tay-Sachs child would involve only those in which both were carriers.

Since we can predict zero risk in every case except when both prospective parents carry the hidden recessive allele, what are the chances of a Tay-Sachs birth in the risk family? Since the gene is autosomal, each parent will produce the same kinds of genetic gametes and, since each is a carrier, there will be two kinds of eggs and two kinds of sperm. Approximately half of the time the egg will have the dominant allele, and half of the time the recessive will be present; and the same chances are true for the sperm produced by the father. Each fertilization is a random event, and any kind of egg may be fertilized by any kind of sperm leading to conception and pregnancy. The chance that an egg has the recessive allele is 1/2, and the chance that the sperm that fertilizes this egg also carries the recessive allele is 1/2; the chance that these two particular kinds of gametes will fuse is $1/2 \times 1/2$, or 1/4. This is the same as asking the chance that two coins tossed at the same time will both land as heads; the chance for each coin to show heads is 1/2, and the chance that both will land as heads is the sum of the separate probabilities, or $1/2 \times 1/2 = 1/4$.

What we have stated is that there is a 25 percent probability, or risk, in *each* pregnancy that the two gametes that unite will happen to carry the same recessive allele. If this does occur, then the fertilized egg will develop into an infant with the recessive allele for the Tay-Sachs gene on each of the two autosomes, one from the father and one from the mother. There is exactly the same 25 percent risk in every pregnancy because the same chances exist for a particular egg to be fertilized by a particular kind of sperm. Because of this, we cannot predict that a couple with one Tay-Sachs child can now expect their next three children to be normal. Neither can we predict that any particular pregnancy or any particular number of children in a family will or will not develop Tay-Sachs disease when both parents are carriers. The same 25 percent risk is predicted for each pregnancy planned by such a couple. When we examine a large sample of a carrier population, however, we will find that approximately one-fourth of the births were of Tay-Sachs infants. But we rarely will find the precise proportions of three normal to one diseased child in a single family. Predictions of risk depend on determining the probability of a certain set of events taking place, and probabilities are not the same as certainties.

Precisely this situation holds true for other genetic traits that develop under the control of the recessive alleles of a single gene; that is, there is a 25 percent probability in each pregnancy that the child will have two recessive alleles of the gene and exhibit the recessive characteristic if both parents are carriers. Sickle-cell anemia also results from the presence of two recessive alleles for the hemoglobin gene that is involved. The defective hemoglobin that is responsible for the sickling disease leads to a variety of painful symptoms and contributes to the premature death of recessive individuals, generally between the ages of 20 and 40. The possibility for a longer life increases every year as improved medical care becomes available to overcome the periodic crises of disease expression. If one or both prospective parents have only normal hemoglobin alleles, then there is no risk in having a child with the sickling disease, since at least one normal allele will be transmitted through the gametes. We already have stated that the risk is 25 percent when both parents are carriers. But unlike the Tay-Sachs disease that causes death in childhood, adults who have the sickling disease can have their own children. The risk of having a child with sickle-cell anemia increases when one parent is recessive, but only if the second parent also is a recessive or carries the allele. If one of the parents is recessive and the other has only normal hemoglobin alleles, there is zero risk of their having a child with the sickling disease. Since the parent with normal hemoglobin alleles contributes one of these in every gamete, every child will have one dominant allele that will mask the recessive that is contributed by the other parent who has the sickling disease. But if one parent is a carrier and the other is recessive, then there is a 50 percent risk in each pregnancy of having a child with sickle-cell anemia. The recessive parent contributes one recessive allele to each child through the gamete, all of which have this allele, and the carrier parent produces two kinds of gametes: half with the dominant and half with the recessive allele. There is thus a 50 percent chance that a gamete carrying the recessive allele will fuse with another from the parent who only contributes recessives. In the event that two people have the sickling disease, we would predict 100 percent risk since every gamete that each parent produces will have one recessive allele and all fertilizations will lead to the presence of two recessive alleles of the gene in the children.

  Both traits that have been described above are produced by recessive autosomal alleles of a single gene. But these alleles, along with many others, do not occur in the same proportions in all human populations. Most but not all families in which Tay-Sachs disease occurs are Ashkenazim Jews, who trace their ancestry to forebears from northeastern Poland and southern Lithuania. The disease is much less common in Sephardic Jews and in non-Jewish people. About 1 in 30 Jews living in New York City is estimated to be a carrier of the Tay-Sachs allele. Sickle-cell anemia occurs in blacks of African descent with the highest frequency, but the genetic condition is not rare in some of the European countries that border the Mediterranean Sea, such as Italy. Among black Americans, about one in 400 to 500 children and adults has the sickling disease, and approximately 9 or 10 percent of the population of 22 million carries the sickle cell trait. This knowledge of the different distributions of particular alleles of the same genes is of great importance in focusing on the most likely genetic problems that certain members of the population might expect to encounter.

Using the same fundamental principles of inheritance, it is possible to predict the probability or risk in cases that result from the action of autosomal dominants or from recessive and dominant sex-linked alleles of the set of genes that we all have in common. The family history provides the framework of reference from which predictions may be made or probabilities calculated for the appearance of genetic afflictions among the children. The exact probabilities will vary depending on the nature of the inheritance pattern and the particular details of an individual family.

Prenatal Diagnosis

For very few genic or chromosomal conditions, it is possible to determine the characteristics of the developing fetus. This option provides the family with information that may permit them to plan more practically for the eventual birth, to relieve or confirm anxieties about the health of the fetus, or to decide whether or not to have the fetus aborted. People who want this kind of information generally have some basis for suspecting the existence of an abnormality in the fetus: either because of a family history of some genetic disorder, a previously afflicted child, a history of spontaneous abortion, or other similar situations. Two genetic abnormalities are particularly suitable for early prenatal diagnosis; one of these is due to a defective allele of a gene and the other to a chromosome anomaly. The principles to be illustrated in these two examples can explain other situations in which suitable features permit the early diagnosis of defects.

Test for Enzyme Activity

Many genic disorders result from an insufficiency or absence of a vital enzyme that is required for metabolism or tissue growth. Children who develop Tay-Sachs disease fail to produce an enzyme that is necessary for processing fats in the diet. In the absence of the enzyme, there is an accumulation of fats, especially around the nerve cells, that leads to their eventual destruction and the death of the child. The human fetus floats in liquid inside a sac formed from an embryonic membrane, known as the *amnion*. It is possible to obtain a sample of cells that are regularly shed by the growing fetus into this liquid, using the technique known as **amniocentesis** (Fig. 3.8). A hypodermic syringe is inserted through the

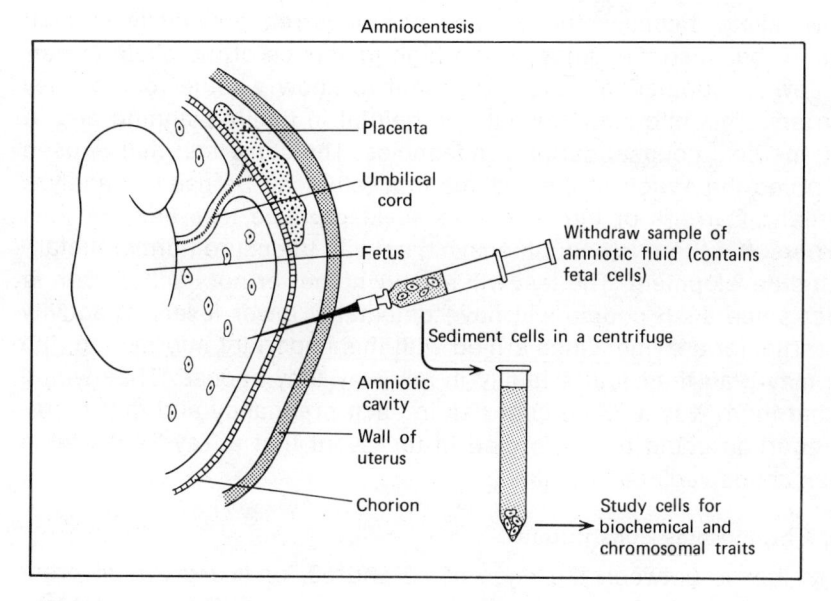

Amniocentesis

figure 3.8
During amniocentesis, a sample of fetal cells and fluids is withdrawn from the amniotic cavity in which the fetus floats while in the uterus of the mother. This sample can be processed in different ways to analyze the chromosome number and appearance, number of Barr bodies in the nucleus, or for biochemical traits in prenatal diagnosis. The hypodermic syringe is inserted through the mother's abdominal wall directly into the amniotic cavity, without touching or damaging the fetus within.

mother's abdomen into the fluid-filled amnion sac that contains the fetus, and a sample of shed fetal cells and fluid is withdrawn for study. If this material gives a negative test for the vital enzyme, then the diagnosis indicates that the fetus will develop the Tay-Sachs disease, which usually happens at about six months after birth. If such an infant is born, he or she will die before the age of four years and have undergone a vegetable existence in the expensive confines of a hospital for most of those years.

The prospective parents may know that they are carriers of the Tay-Sachs allele if a previous child developed the disease, in which case they may wish to know the condition of an expected child during the early months of pregnancy. They can learn whether or not the new fetus is recessive for Tay-Sachs by amniocentesis and the subsequent analysis of the fetal cells and fluid. If the disease is indicated in the fetus, then the parents may decide on abortion, or prepare themselves for another heartbreaking experience.

If prospective parents have reason to believe that each may carry the recessive allele, because the disease has occurred previously in their families, or because they know of the high incidence of the allele in particular Jewish populations, they may want to know whether or not they are carriers. This information would be helpful in family planning and in seeking medical counsel during pregnancies. The same test that is used to determine the action of the enzyme in a fetus can be used to analyze the parents. Carriers of the recessive allele produce less enzyme than noncarriers, but there is enough enzyme activity to ensure normal metabolism and development. The test will show whether or not each person is a carrier since such people will have detectably lower levels of activity of the particular enzyme. Once armed with this important information, the couple may plan their future family in any way they choose. They would know that there was a 25 percent risk in each pregnancy and that therapeutic abortion could be performed in the event that a Tay-Sachs fetus has been conceived at a later time.

Test for Chromosome Constitution

If the mother is between the ages of 40 and 50, or if a previous pregnancy has resulted in a child with Down's syndrome, the parents may wish to know if the developing fetus in the pregnant woman has this condition. A sample of fetal cells is removed by amniocentesis and the cells are sedimented to the bottom of a tube. These cells then are prepared for examination with the microscope. If there are 47 chromosomes and the extra one is chromosome-21, then Down's syndrome is indicated. Then, the prospective parents can decide on their subsequent course of action: abortion or birth of the mongoloid child.

In some cases of sex-linked disorders, the parents may wish to know if the developing fetus is male or female since such genetic traits usually occur in males and very rarely occur in females if the inheritance pattern is that of a sex-linked recessive allele. Once again, a sample of fetal cells would be removed, prepared, and examined with the microscope to see if the sex chromosomes are **XX** or **XY**. A much simpler test would be to stain the cells withdrawn by amniocentesis and see whether or not a Barr body is present in the nucleus. One Barr body would indicate a female fetus while male cells would lack the Barr body.

Genetic Counseling

Far too little information is available to the general public, and all too few physicians have an adequate working knowledge of medical genetics to provide an appropriate level of health care and counseling. Medical schools have introduced greater amounts of genetics into their curricula in recent years, but at an astonishingly slow rate. This is unfortunate since the genetic counseling team cannot function without the primary input of the diagnostic physician.

In addition to the physician, whose province is diagnosis and rehabilitative treatment, other members of the counseling group would include professionals who were trained in human genetics, laboratory personnel who perform and may interpret tests, and the professional counseling aides. Counseling aides participate in the pedigree studies, counseling of the family and in the follow-through over a period of time, and provide various other kinds of expertise required for successful team efforts. Genetic counseling is available in some parts of the United States through state-subsidized genetics clinics, in others through modest private resources, and not at all in some parts of the country. The twin problems of too few trained personnel and the usual minimum funding for health care programs and facilities have been responsible for these gaps in health care and education. Among the best of the genetics clinics and counseling centers are those associated with departments of Medical Genetics and Pediatrics in medical schools and their affiliated hospital centers.

Along with the training of physicians with an adequate background in medical genetics, there has been an increase in professional people with a Ph.D. degree in human genetics. The latter group provides the new and updated information from population studies and basic research programs. Two schools presently offer a Master's degree program to train genetic counseling aides: Sarah Lawrence College, which initiated a program in 1969, and Rutgers University, which began the graduate program in 1972.

Suggested Readings

Allison, A. C., Sickle cells and evolution. *Scientific American, 195* (August 1956), 87.

Bearn, A. G., The chemistry of hereditary disease. *Scientific American, 195* (December 1956), 126.

Bearn, A. G., and J. L. German III, Chromosomes and disease. *Scientific American, 205* (November 1961), 66.

Cavalli-Sforza, L. L., "Genetic drift" in an Italian population. *Scientific American, 221* (August 1969), 30.

Friedmann, T., Prenatal diagnosis of genetic disease. *Scientific American, 225* (November 1971), 34.

German, J. L., Studying human chromosomes today. *Amer. Scientist, 58* (1970), 182.

Hannah-Alava, A., Genetic mosaics. *Scientific American, 202* (May 1960), 118.

Ingalls, T. H., Mongolism. *Scientific American, 186* (February 1952), 60.

Lubs, H. A., and F. H. Ruddle, Chromosomal abnormalities in the human population: Estimation of rates based on New Haven newborn study. *Science, 169* (1970), 495.

McKusick, V. A., The mapping of human chromosomes. *Scientific American, 224* (April 1971), 104.

Mittwoch, U., Sex differences in cells. *Scientific American, 209* (July 1963), 54.

Winchester, A. M., *Human Genetics*. Columbus, Ohio: Charles E. Merrell Publ., 1971.

chapter four

Evolutionary Considerations

Life on the Earth probably originated here more than 3 billion years ago, at least one billion years after the planet itself assumed its final form 4.6 billion years ago. A long period of chemical evolution preceded the appearance of the first life forms, and the evolutionary processes that shaped living system were quite different from those that influenced prelife changes. Living systems have undergone a continuous sequence of modifications since those ancient origins, and continue at this moment to be reshaped in quantity and quality. Before we concentrate on the evolutionary themes directly related to sex, we must have a general basis for understanding and interpreting the particular changes that did occur. The fundamental principles that govern biological evolution are essentially the same for all forms of life and for all sorts of changes that bind together the world of living things.

Some Evolutionary Principles

Any casual inspection of the smallest segment of life on the Earth immediately reveals an incredible spectrum of variation. There are plants, animals, microbes, and viruses; and aquatic, terrestrial, and flying forms, large and small, long-lived and short-lived, in a seemingly endless series.

GENETICS AND EVOLUTION

Evolution can be described as *descent with modification*, which means that the life we see around us now can trace its history back through billions of years of time; that each form is genetically different from its remote ancestors, but retains a common heritage with closer relatives. How did all of this diversity of life arise? Once produced, how is this diversity incorporated into the successful species that maintain the continuity of life on our planet?

Origins of Diversity

Mutation

The only process that is known to produce totally *new* genetic information is **mutation**. The mutated, or altered, coded information in the DNA molecules leads to a different set of proteins from those that are manufactured according to instructions in the predominant (wild type) alleles of the same genes. Different kinds of changes in DNA can be considered mutations, if the changes appeared suddenly and were inherited in later generations. Since the mutant and wild type alleles of the same gene will be transmitted to the progeny, there is a continuity of the genes and therefore of the species. The new mutant information may spread through later generations and populations if it is transmitted to increasing numbers of individuals. We expect that any mutation that is harmful will spread very slowly, if at all, whereas a new beneficial mutation may multiply rapidly in a few generations and perhaps come to be characteristic of most of the members of a population, a race, or a species.

The rate of mutation is different for different genes, different parts of the same gene, and different kinds of life. Since each mutation occurs by chance, there is a characteristic probability for a particular mutation to occur. We can't predict when or where the mutation will happen, but we can predict how often it will happen on the average. There is no known method by which we can direct a particular mutation to occur in a particular gene. But there are physical agents such as x-rays and many chemicals that cause an overall increase in the rate of gene mutations. These agents increase the probability that the DNA will undergo an alteration, but their effects are not specific for any particular gene.

Some mutations are beneficial to a population under one set of living conditions but may have no advantage in other places or at other times. The recessive allele for sickle-cell anemia may occur in a few or no members of some populations, or in as many as 40 percent of other communities in Africa. Since the sickling disease causes premature death, why should such an allele be so common in some areas? It is known that carriers of the recessive allele essentially are resistant to the deadly malarial infections that are common in some areas, whereas people lacking the allele are sensitive to the blood parasite. There is a clear advantage in malaria-infested regions for those people who can resist the infection. We believe that the high percentage of carriers is due to this particular advantage but, when malaria is not present, this same allele would remain rare rather than spread through a population because of its otherwise harmful nature. In the United States population of blacks, the proportion of carriers of this trait has been reduced to 9 or 10 percent from the estimated 22 percent of two or three centuries ago. This percentage should continue to decline in future generations.

The gradual incorporation of beneficial mutations leads to changes in populations and to new species and forms of life. Since mutations are random events, we would not expect the same genes to mutate in every population or even in closely related species. Different genes mutate at different times and in different populations, all of which leads to an increasing level of genetic variation and to the diversity of life during the eons of evolutionary time.

Recombination of Genes

The rearrangement of existing alleles into new combinations, or **recombination**, does not produce new genetic information, but we believe that this phenomenon is crucial for rapid and meaningful increase in population diversity. The significant feature of recombination is the *speed* with which new combinations of alleles may arise and be spread within and between populations. An example may help to illustrate this point. Suppose we had a population in which different individuals were resistant to infections of three different kinds, but any one organism could resist only one of the three diseases and not the other two. By chance mutations, some occasional individuals might become resistant to a second infection and, very rarely, even to all three. As long as the diseases did not afflict the population all at the same time, some individuals always would survive while others might succumb to one of the infections. If additional resistance could arise only by chance mutation in occasional individuals, there would be none or very few who could survive a combination of infections. This population might be well adapted to the usual living conditions, but if these conditions changed so that all three diseases occurred at the same time, it would lead to very few survivors if any. The incorporation and spread of the resistance alleles is very slow when mutation is the only mechanism for producing new combinations. If, on the other hand, there was a mechanism for exchanging and recombining genes from different individuals, then more of the population could inherit the most favorable combinations of all three resistance alleles. Recombination not only would lead to a greater proportion of resistant individuals in the population, it also permits this change to take place in one or a few generations. At any time that all three diseases afflicted this population at the same time, more of the population would survive, and there would be a relative increase in this more successful genetic type as the others failed to survive the multiple infections.

The generation of new combinations of alleles of the existing genes is one of the most important aspects of evolutionary success. The capacity for gene exchange between individuals is a constant feature of sexual reproduction. The primary adaptive advantage of sexual reproduction is the *high frequency* and the *regularity* with which new combinations of alleles can be produced. The numerous improvements that have marked the evolution of sexual systems are directly related to the greater regularity and efficiency in producing gene recombination. These improvements include a higher degree of certainty that gametes from different genetic backgrounds will fuse, thus increasing the genetic diversity in populations in every generation.

Changes in Chromosomes

In addition to diversity that results from varieties of alleles of the genes, different genetic types may arise in populations when there are changes in the number of chromosomes in the cell and when parts of chromosomes undergo some structural reorganization.

There are four general kinds of structural rearrangement of chromosomes, all of which require the prior breakage and subsequent reunion of broken ends in different patterns from the original. If some part of a chromosome is lost, or *deleted*, there often is the loss of genes that are vital to the normal development of the organism. Deletions often mimic the effects of recessive alleles, especially those that lead to absent or defective proteins. In some situations, it may be an advantage for a population to lose some genes, but we expect that the usual effects of deletions are harmful to the species. Deletions thus would contribute to a lower chance of evolutionary success for a species, but we really know almost nothing about the long-term consequences of this kind of chromosome change.

If chromosomes break and then rejoin so that a piece of a chromosome is present in duplicate or more copies, then additional genes have been added to the chromosome sets that occur in the nucleus. At one time we had very little information about such chromosome *duplications*, but with new methods for studying the DNA we have discovered that most species contain a very high amount of repeated genes. There have been some suggestions to explain the significance of so much duplicated genetic material, but we have too little evidence at present to choose among these suggestions. One of the ideas that has been especially favored is that the higher forms of life actually contain relatively few *kinds* of genes, perhaps thousands, which direct the synthesis of the proteins needed for metabolism and growth. All the rest of the DNA then may be repeated information that regulates or controls the action of these few thousand primary genes of the species. Since more complex forms of life have essentially the same sorts of proteins as we find in simpler organisms, we would not expect more genes to be involved in one kind of life than in another. But there is a profound difference in the complexity of the *development* of structures, functions, and behaviors in higher forms of life; and perhaps the developmental complexities have evolved successfully because repeated DNA exerts a fine level of control in directing these processes of growth and differentiation. This is an extremely interesting and important concept of genetic evolution, and it may provide significant clues that will increase our understanding of the differences and similarities among the many kinds of life on the Earth.

There are two other chromosome break-and-rejoin phenomena that have been studied intensively for many years, and each has been shown to have an important effect on species evolution. In one case, a chromosome may break in two places and, when this piece is reinserted into the same chromosome, it may have become inverted 180° from its original orientation. This *inversion* results in a different sequence of the genes on the chromosome, which presents no problems if the partner chromosome also contains this same inversion. When the pair of chromosomes is different and only one of the two is inverted, then problems occur during meiosis and the gametes that are produced often are unable to function because of gene unbalance. Chromosome inversions are often responsible for the sterility of a hybrid between individuals whose genes are in different arrangements. Even if gametes fuse and the fertilized egg does develop into a hybrid individual, the hybrid itself may not be able to produce functional gametes because of the unbalanced genes that are produced after meiosis. Such differences provide one of the mechanisms that keep species separate, by the failure of hybrids to reproduce. Once species cannot interbreed, they undergo their individual evolutionary changes and do not share new genes or gene recombinations. This genetic basis for preventing interbreeding is one of many that keeps species on their separate evolutionary tracks and contributes to the preservation of diversity.

The breakage and rejoining of broken chromosomes may involve two different chromosomes, unlike the one chromosome involved in an inversion. When chromosome pieces from different sources reunite, then there is a scrambling of the genes, but not necessarily any loss of genes. This phenomenon also leads to hybrids that are sterile because the scrambled and unscrambled chromosomes do not function together at meiosis, and the gametes are not functional. This *translocation* of genes from one chromosome to another also provides a basis for understanding why hybrids between different species do not continue to reproduce more hybrids and merge the genes of two species into one common pool. Species with different gene arrangements on their chromosomes remain separate from each other, and each evolves independently of the other.

Changes in the number of chromosomes probably have been very important during evolution, since additional genes are added to a species in this way. The genes on the extra chromosomes would persist only if there were no problems of gene unbalance to cause developmental abnormalities. If you recall, the presence of extra autosomes or sex chromosomes in humans led to numerous difficulties for the individuals concerned. But in some species we know that extra chromosomes do no damage, and all we can propose is that added chromosomes may have been important in the evolution of some species and not of others. More important than adding a chromosome or two is the doubling of the chromosome number, or chromosome sets, in various multiples. Doubling the chromosome number from one set to two sets in the adults of the species and retaining the condition of only one set for the gamete stage was one of the most significant evolutionary inventions. We will discuss this shortly when we examine the evolutionary steps that have contributed to sexual reproduction and its improvements with time.

Some of our most important crop plants have more sets of chromosomes than did their wild ancestors. Tobacco grown for profit is characterized by four sets of chromosomes, although the wild species have only the usual two sets. Wheat species that are used most widely for human foods have six sets of chromosomes, but some types have four in contrast with the two sets in the wild species or the most ancient types that still are of agricultural importance in some parts of the world. There is a long list of important plants with more than two sets of chromosomes in addition to tobacco and wheat, including potato, strawberry, and many others. There are many fewer difficulties in plants with extra sets of chromosomes, and about 50 percent of the known flowering plants are of this type. Animals usually have only the standard two chromosome sets, which is understandable in view of the delicacy of the sex chromosome mechanism for sex determination. Extra sets of autosomes and sex chromosomes might lead to great variations in the chromosome number that is distributed to the gametes and unbalanced chromosome complements in the individual animals. The abundance of extra sets of chromosomes in plant species and the rarity in animals, especially vertebrates, almost certainly reflects the properties of sex chromosome mechanisms for sex determination.

Natural Selection and Adaptation

We have indicated the various mechanisms that increase species diversity, but it is obvious that many gene or chromosome changes are harmful and that only some are beneficial to the population. The steady improvement in life forms throughout evolutionary time would seem to be very unlikely in view of the many more frequent disadvantageous gene changes that do occur. How do species continue to improve? Why do harmful alleles remain relatively rare in populations even though mutations are known to produce some harmful effects? How do the rare beneficial gene changes become more common and spread throughout populations leading eventually to entirely new forms of life? The simple but profound process that provides the answers to these questions is that of **natural selection.** This concept first was described in 1859 by Charles Darwin in *The Origin of Species*, a book that raised a storm of controversy in his time, and continues to provoke differences of opinion to the present day among some groups of people. The concept of natural selection leading to adaptive change in all forms of life stands as a central unifying theme in biology.

The definition that is often used to describe natural selection as a process is: *differential reproduction of genetically different populations.* All this means is that diversity exists and, among this diversity, there are some types that will leave more progeny and thus perpetuate their own lineage as well as increase in numbers and in proportion to other genetic types in their surroundings. It is not "nature red in tooth and claw" or a "struggle" for survival in the sense of bloody fights over mates and resources. The inherent advantages of one genetic type over another make it more likely to reproduce and to perpetuate those same genes. The greater disadvantage of another genetic type makes it less likely to reproduce, and leave fewer progeny, hence fewer of its kinds of genes. Eventually, the more adapted forms will predominate in a population and that population will then be different from its predecessors. This continuing process of differential reproduction, some types leave more progeny than others, provides much of the answer to the progression of improvements in life on this planet.

An alternative but unproven explanation for continued improvements during evolution would be that there was some general "plan" of evolution, or that there was an inherent "drive" toward improvement. Although there are people who believe in philosophical or metaphysical forces, the actual hard evidence shows overwhelmingly that evolution is the result of random events without plan or ultimate goal. The random gene and chromosome changes are perpetuated if they are beneficial, and lost or remain rare if they are detrimental. The apparent direction toward ever-increasing improvement is the result of the incorporation of adaptive inherited changes, under the influence of natural selection (Fig. 4.1). We cannot predict future evolutionary innovation, nor could we have predicted the actual events of evolution if we had been observers in the ancient past. The theory of natural selection has provided a unitary basis by which many life phenomena have been and can be explained, and it stands as one of the major intellectual achievements in our history.

The Evolution of Sex

Even the most casual survey of life forms reveals that the most ancient life forms are primarily or exclusively asexual and that sexual reproduction evolved from some ancestral asexual mode. It is also true that some highly evolved species, especially among the flowering plants, reproduce by some asexual method. There is excellent evidence that such asexuality arose very recently in evolution by the loss of sexual capacity. One or more fairly simple genetic modifications could lead to the loss of sexuality, but the acquisition of a sexual system would require a series of related genetic changes, and their incorporation in stepwise fashion, under the influence of natural selection throughout the sequence.

The crucial events of sexual reproduction are meiosis and fertilization, each process being under complex genetic control. The origins of these processes are unknown, however. Meiosis and fertilization alternate during the sexual cycle, and a constant number of chromosomes and genes is maintained for a species. The chromosome number is doubled when gametes fuse, but the chromosome number in the gamete previously was reduced to one-half during meiosis. There would be genetic chaos otherwise, so that every successful sexual system must include these two processes coordinated and integrated into a common program.

Evolution of Gametes

The gametes, or cells that are capable of fusion to produce a new individual of the species, may be identical in size and form, chemistry, and behavior; or they may be widely different from each other. Many of the simpler kinds of life produce gametes that seem to be identical, or at least indistinguishable. These sex cells may originate from ordinary body cells or in specialized sex organs, with much variety between these

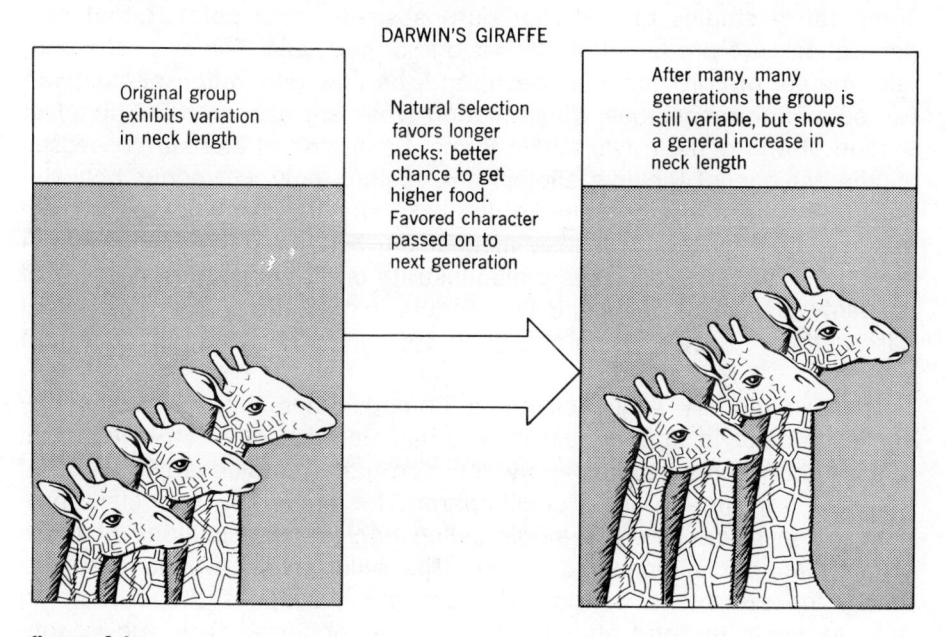

figure 4.1
The evolution of improved or more advantageous traits has been explained in various ways at different times in history. In 1809, Jean Lamarck proposed the Theory of the Inheritance of Acquired Characteristics, and he discussed the lengthening of the giraffe's neck as an example of the action of this process. There is no evidence to support this theory, and it is not accepted by the scientific community. The explanation put forth by Charles Darwin, in 1859, proposes that the same evolutionary event would have occurred by natural selection of random variations and not by directed modifications.

two extremes. In some lower species, the same cell may reproduce asexually at one time and then assume the behavior of a gamete under another set of conditions.

Incompatibility and Gamete Differences

The most common situation in sexual species is the one in which there are two kinds of gametes, although some systems have more than two kinds. The gametes may belong to different compatibility groups, mating types, or sexes. These differences in description reflect the observation that we are not always able to recognize eggs and sperm or sex cells that look or behave differently from one another. In many of the lower plant and animal species, we only know that there are different mating types because cells will not fuse within a strain, but only with cells from a genetically different population of the species. We can call these mating types or indicate that the strains belong to different compatibility groups, or call them sexes. The terminology may be of importance in detailed comparative studies but, for our purposes, the main point is that new generations are produced by the fusions of sex cells. The sex cells contain one set of chromosomes each, and the new generation begins with two sets of chromosomes. Some mating types are based on the simplest system of one gene in two allelic forms. Each kind of gamete possesses a different one of the two alleles, and fusions only will occur between cells that have a different allele of the sex gene. Sexuality could not have arisen simply by this one-gene mechanism because other genes direct meiosis, at the least. But this does illustrate one of the simplest forms of sexuality, based on one gene that occurs in two alternatives. Cells that carry different alleles of the one gene are compatible and may fuse in a sexual act.

The evolutionary sequence leads from gametes that are essentially similar in behavior and appearance to gametes that differ strikingly from one another. The overwhelmingly common pattern includes two kinds of gametes, a large egg and a small sperm. The disparity is especially pronounced in birds, where a single-celled egg is many millions of times larger than the microscopic sperm. The enlargement of the egg permitted the advantage of food storage, which sustains the developing embryo within the egg after fertilization has occurred. The egg is non-motile, although the sperm swim to reach the egg. The usual pattern involves the production of many more sperm than eggs, which is an obvious adaptive feature in view of the more hazardous existence of the motile sperm. The essential ingredient of the gamete is its set of genes, and the egg and sperm provide equivalent genetic contributions to the next generation. The increase in size of the egg reflects its nutrient stores, and this factor also explains its lack of movement relative to the swimming sperm.

Separation of the Sexes

The incompatibility of sex cells of the same mating type guarantees that two genetically different cells will initiate the new generation. But there is no assurance of genetic difference between egg and sperm if they are produced by the same individual. Organisms that produce both eggs and sperm in the same individual are called **hermaphrodites** (after the Greek gods Hermes and Aphrodite). The evolution of hermaphrodites probably occurred by the differentiation of sex organs in which one developed the potential only for egg production and the other only for sperm. This kind of tissue specialization is similar to differentiation of the heart, liver, skin, and other structures that mature into differently functioning cell systems in one individual; but there must be the same genes in all of these cells. Once again, there is the regulation of gene expression that produces different end results, depending on which genes are turned on and which are turned off during development.

Among hermaphroditic species, such as the earthworm and many flowering plants, there is very little self-fertilization. The commoner arrangement is for one individual to deliver sperm to another hermaphrodite and also to receive its sperm in the sex act. This cross-fertilizing pattern makes it more likely that genetically different gametes will unite, thus increasing diversity in the species populations.

An evolutionary improvement on the hermaphroditic pattern involves the separation of the egg-producing and sperm-producing organs into different individuals. In this way, there is an ensured genetic difference between the two gametes and a greater variability among the fusion products that develop into the adults of the species. The adaptive significance of separate sexes resides in the greater probability, or the certainty, that gene recombination will take place in each fertilization event. Cross-fertilizing species have a high level of genetic variability, which is maintained by continued interbreeding between the genetically different individuals in the population.

There is some evidence from experimental studies of hermaphroditic species, such as the corn plant, that relatively simple gene changes may lead to a separation of the sexes into different individuals without a disturbance in the gene balance of the population. Similar genetic events may have led to the evolution of separated sexes from the more ancient hermaphroditic pattern that exists in both plant and animal groups.

Haploid to Diploid

In each sexual cycle, there is the process of fertilization that brings together two sets of chromosomes, initiating the **diploid** phase; and meiosis, which reduces the number to one set of chromosomes, or the **haploid** phase. Many of the present-day lower forms of organisms exist primarily in the haploid phase and are only transiently diploid. In these systems, the organism is initiated at fertilization, but the cell almost immediately undergoes meiosis, and the haploid phase is restored. All multicellular, tissue-forming animals and most of the higher land plants exist in the diploid phase for almost the entire life cycle. For animals, the only haploid cells are the gametes themselves, which live for a brief time and die unless sexual fusion occurs. The evolutionary sequence that we interpret from observations of present-day life forms involved the gradual extension of the diploid phase and the simultaneous shortening of the haploid phase of the life cycle until only the gametes remain haploid.

The advantage of diploidy is that a great deal of genetic variability can be stored in the population and not be expressed immediately. Haploids have only one set of genes in the one set of chromosomes so that every allele would be expressed, whether harmful or not. Diploids have two sets of alleles so that recessives may be masked from expression if dominant, normal alleles exist side by side in the cells with these recessives. The stored variability is released slowly and gradually in sexual diploid species, as gene recombinations lead to different kinds of individuals in a population. A diploid population thus retains much of its evolutionary potential and is genetically more flexible in meeting the challenges of new or fluctuating environments. The longer the interval of diploidy in the life cycle, the more protected the individual from the effects of the relatively more harmful recessive alleles, at any one moment in time. It is the species as a whole that is more likely to survive many kinds of challenge if its genetic reserves are adequate. Diploidy provides a high level of insurance that these reserves will be present and that some members of the population may be suited to a new living condition, if this happens to develop at some time in evolution. We each carry hundreds of recessive alleles, many of which are potentially harmful when expressed. About 1 person in 20 carries the recessive allele for cystic fibrosis, 1 black in 9 or 10 carries the sickle-cell allele, and 1 Jew in every 30 living in New York City carries the Tay-Sachs allele; yet relatively few people express these traits because the number of diploid recessive individuals is much lower. Of course, not every recessive allele is harmful. Those occasional beneficial recessive alleles may provide the edge for survival of a species, if the expression of the trait permits at least some reproduction and the perpetuation of a species under conditions of change and challenge.

Protection of the Egg and the Embryo

During vertebrate evolution, numerous improvements became incorporated into populations, among which were the greater protection of the egg and the embryo from external hazards. Except for some fish that bear live young, the eggs of most species of fish are released into the water and are fertilized outside the body by sperm that are shed from the male. Elaborate patterns of courtship and mating behavior have evolved, some of which almost guarantee that the right kind of sperm will be present at the right time to fertilize the eggs. But once these eggs have been fertilized, some embryonic development takes place using stored foods in the egg. Once they are hatched, the young fish are on their own, and relatively few survive their first rigorous minutes or hours in the outside world.

The land vertebrates show a great deal of variation in the level of protection that is afforded to the eggs and embryos. Amphibians such as frogs, toads, and salamanders release eggs and sperm into the water, where fertilization takes place. The embryo develops to an incomplete larval stage before the egg hatches, and the hatchling continues its growth and development to the adult form if it successfully overcomes predators and other hostile elements in the water. Water is essential for fertilization in amphibians so that they depend on this medium for reproduction and continuity. Even in the amphibians that spend more time on the land, such as some frogs and toads, there is an absolute requirement for water to accomplish sexual reproduction. The amphibians and their fish ancestors differ in many ways, but each must live near water if the species is to survive through time.

One of the many significant novel improvements that permitted the reptiles to be free from a dependence on water for reproduction, unlike their amphibian ancestors, was the evolution of an egg that is retained within the female. This permits internal fertilization and requires many modifications in both sexes if the male is to deposit his sperm within the female tract. Except for some modern reptiles that bear their young alive, especially among the snakes, the developing fertilized egg is expelled from the mother before hatching. The eggs continue to furnish food and shelter to the developing embryo within. For this to be possible without drying out, there must be a protective covering which prevents the desiccation of the egg. The success of the reptilian egg thus depended on the evolutionary development of a tough, leathery covering, as well as retention within the female during and after fertilization. When it hatches, the newborn resembles the adult form rather than an undeveloped larva like the frog tadpole. The larger size of the egg and the rich store of nutrients permit a lengthier interval of embryonic development. Some aquatic reptiles have evolved more recently from ancestors that were land reptiles. Crocodiles, alligators, and sea turtles return to the land to lay their eggs, whereas some species of sea snakes bear their young alive and never spend time on the land, even to reproduce.

Birds and mammals both are descended from reptilian ancestry, but birds have retained many reptilian features. They sometimes are called "reptiles with feathers," but even the feather is a modified reptilian scale. Fertilization is internal in birds and mammals, as in the reptiles, but all birds lay eggs so that there is an interval of embryonic development outside of the mother's body, until hatching. Young birds resemble the adult, but they require a period of parental care before they can assume their independence. Except for the egg-laying platypus and spiny anteaters of Australia and New Guinea, all other mammals bear their young alive, after a period of development (*gestation*) in the mother's uterus. The marsupial mammals such as kangaroos and opossums begin life in the uterus but, shortly afterward, the young embryos crawl out of the uterus to the external pouch on the mother's abdomen. They immedi-

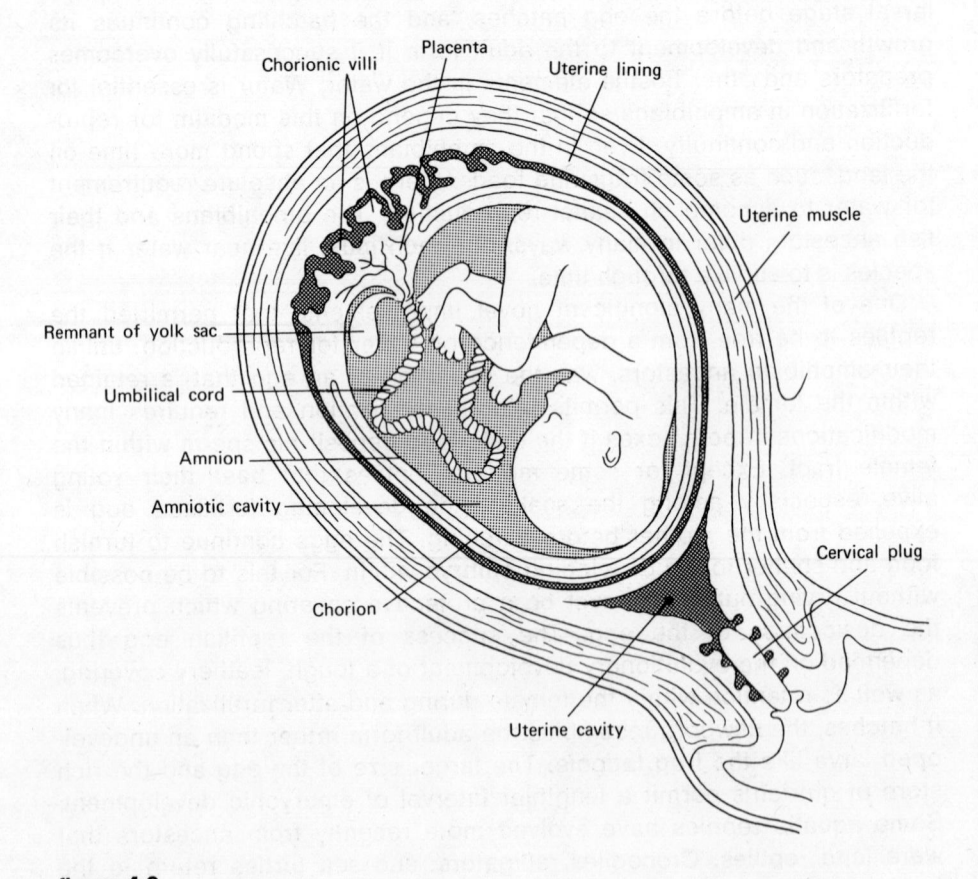

figure 4.2
The mammalian embryo develops within membranes in the mother's uterus and receives nutrients and eliminates wastes through the umbilical cord and the fetal-maternal tissues of the placenta.

ately clamp their jaws on one of the teats and remain attached there for most of their subsequent development. The muscular efforts of the mother literally pump milk through the teat to the young, developing *fetus*. Until it becomes totally independent, the young marsupial remains in its mother's pouch even though it leaves for short periods much later in its development.

The placental-mammal embryo develops within the maternal uterus for a relatively longer developmental interval, and it is connected to the maternal circulation via the umbilical cord and placenta (Fig. 4.2). The placenta develops from a combination of maternal and fetal tissues, and represents one mammalian innovation that permits the developing fetus to receive nourishment and to eliminate wastes long after the nutrient supply in the fertilized egg has been exhausted. The evolution of the placenta provided a mechanism that allows a longer interval of embryonic and fetal development within the protection of the mother's body. Some newborn mammals can see, hear, and run within hours after birth; others require a period of intensive care after birth since they enter the world blind, deaf, and essentially helpless. All mammals, whether egg-laying, marsupial, or placental, nourish their young with milk secreted by the mother. The higher primates, such as monkeys, apes, and humans, bear helpless young that require a long period of parental care. But this interval also provides a period of time for learning, socialization, and acculturation so that the young primate enters adult society with some knowledge of the interrelationships that are essential for it to become successfully integrated and a full participant in the social group.

Sex Determination

Systems of Determination

Sex determination in hermaphroditic species cannot be genetic since both kinds of sex organs regularly develop in the same individual. In some simpler animal forms, the determination of sex depends entirely on the nature of the environment in which the undeveloped animal occurs. There is a frequently cited example of a species of marine worm in which the larvae are free-swimming and sexually neutral. If a larva happens to settle on the sea bottom, it will differentiate into a relatively large female; if the larva happens to settle instead on an adult female, then it enters her body, differentiates into a minute male, and leads a parasitic existence there in return for its sperm.

A recent example of reversible sex determination was reported for a fish of the Australian coral reefs. These animals live in small groups that include one male and a harem of three to six females. The male is very aggressive and dominates all of the females in the group, but there is a noticeable social order among the females themselves, and one usually dominates the others. When the male dies, the dominant female undergoes rapid sex reversal and replaces the dead male as the new master of the harem. Each fish possesses both sperm-producing and egg-producing tissues in the sex organs, but one of these two types is suppressed from developing according to the social control system that regulates the sex expression of each individual. Once the dominant female has begun sex reversal, it may take only 14 to 18 days before functional sperm are released and the animal assumes its social and reproductive role as a male. Hormonal influences clearly underlie the changes that are initiated by the social behavior of the dominant members of the group, and males are produced only as needed.

Although more vulnerable systems of sex determination seem to occur in fish generally, the reptiles, birds, and mammals possess a chromosomal mechanism that regulates sex determination. Reptiles and birds have **XY** females and **XX** males, unlike the **XX** female—**XY** male system in mammals. There have been some suggestions concerning the origin of this reversal of identity during the evolution of mammals, but we really have very little solid evidence.

Numerous sex-determining mechanisms have been reported for insects (Table 4.1). In bees and wasps, the males develop from unfertilized eggs and thus have only one chromosome set; females develop from fertilized eggs. Some insect species have more than two sex chromosomes per cell. In grasshoppers, there is only the **X** chromosome in most species,

table 4.1
Summary of the Major Chromosomal Sex Determination Patterns

Males	Females	Species in Which These Occur
XY	XX	Mammals, Drosophilia, some flowering plants
XO	XX	Most grasshoppers, many Hemiptera (bees, ants, wasps)[a]
XX	XY	Some fish, reptiles, birds, and Lepidoptera (moths and butterflies)
Y	X	Liverworts[b]

[a] In many hemipterans, males develop from unfertilized eggs and are haploid adults; females are diploid.

[b] The gamete-producing plant is haploid and contains only one set of autosomes and one of the sex chromosomes.

and the presence of one **X** determines male development while two **X** chromosomes lead to female differentiation.

Intactness of the Sex Chromosomes

Once there was a chromosomal mechanism for sex determination, there would be great adaptive value in making the **X** and **Y** as different as possible from each other in their gene content. If there were sex-determining genes on these chromosomes, then all sorts of unbalance would result if these genes became scrambled because of chromosome exchanges. The system would not function properly from generation to generation unless there was a genetic differentiation such that the two kinds of chromosomes each retained their set of genes in unchanged arrangements.

The **X** and **Y** chromosomes in humans have no genes in common, as far as we know. There is no pattern of inheritance for a trait that indicates its presence on both the **X** and **Y**, but we know so few **X**-linked patterns that we may have failed to detect some. It seems unlikely because gene recombinations between the **X** and **Y** chromosomes would indicate the possibility of defects in the sex determination system, and we have no evidence that this is actually the case in the thousands of studies that have been made.

The sex determination system in mammals thus enforces the production of two sexes, in essentially equal numbers, and insures that genetically different gametes will fuse to produce each new generation. The great success of mammals during evolution is due, in no small measure, to these patterns of enhancing diversity, and thus of insuring that some adaptive types will appear in populations throughout the existence of the species.

Suggested Readings

Avers, C. J., *Evolution*. New York: Harper & Row, 1974.

Crew, F. A. E., *Sex Determination*, 4th ed. New York: Dover, 1965.

Dobzhansky, Th., The genetic basis of evolution. *Scientific American, 182* (January 1950), 32.

Eiseley, L. C., Charles Darwin. *Scientific American, 194* (February 1956), 62.

Kettlewell, H. B. D., Darwin's missing evidence. *Scientific American, 200* (March 1959), 48.

Merrell, D. J., *Evolution and Genetics*. New York: Holt, Rinehart, Winston, 1962.

Robertson, D. R., Social control of sex reversal in a coral-reef fish. *Science, 177* (1972), 1007.

part two
HUMAN REPRODUCTION

chapter five

Reproductive Anatomy and Physiology

General Features

Sexual reproduction is characteristic of all the vertebrate animals, from the many fishes, sharks, and ancient aquatic forms to the most highly developed mammals. The sex cells are produced in individual male and female animals, and once the egg and sperm have united, the first cell of the new generation has been initiated. The fertilized egg undergoes a period of embryonic growth and development, until hatching or birth. Most vertebrate hatchlings resemble the adult, but in some groups there is a period of posthatching development that finally leads to the adult. We are all familiar with the tadpole and the frog; the tadpole hatches from the egg after a period of embryonic development, but it continues to differentiate and finally assumes the adult form of the species. The tadpole is a larval form in the sequence of development from fertilized egg to adult.

Except for some fish that produce live young, all of the aquatic groups and amphibians, such as the frogs and toads, are characterized by external fertilization. Eggs shed by females are fertilized by sperm shed by males, usually in a pattern that leads both kinds of sex cells to be in the same small area. The reptiles were the first land vertebrates that were totally free from a dependence on water to accomplish fertilization.

The reptiles and birds experience internal fertilization, and the embryo develops within the mother's body for a time; but the egg is laid so that further development of the embryo is external to the mother. Some reptiles bear live young and are exceptions in this case. A whole series of related adaptations was required for successful terrestrial reproduction, including deposition of sperm by the male directly into the female reproductive tract, sufficient nutrition and temperature control for the fertilized egg to undergo a substantial interval of embryonic development in its own watery environment, and many other features. Since the embryo continues to develop for a time after the egg is laid, it must be able to withstand the drying conditions on the land and incubate at a proper temperature until hatching occurs. The tough, leathery, or brittle covering of the reptilian and avian eggs prevents desiccation; and deposition of the eggs in a warm nest dug in the ground or in a nest warmed by the parent(s) provides the right temperature during incubation. When the egg hatches, the embryo has completed a large part of its organ development and it physically resembles its parents, in all of the land vertebrates except the amphibians.

The mammals as a group share a number of basic characteristics, one of which is **suckling** their young on milk from mammary glands of the mother. This is true for the egg-laying mammals (platypus and spiny anteater) as well as for marsupial and placental groups of mammals. After hatching, the young platypus suckle on milk that oozes from modified sweat glands (related to true mammary glands) of the mother; young marsupials, such as the kangaroo, have milk literally pumped into them by maternal muscular efforts while the young continue to grow and develop within the external pouch of the mother; and young placental mammals receive milk by a complex process that involves the stimulus of the suckling itself to induce the flow of milk from the mother's breast through the nipples.

The embryo of the egg-laying mammals still is contained within the egg when it is released from the mother's body; marsupial embryos are quite undeveloped at the time they crawl from the uterus to the outside and enter the mother's pouch, and they undergo a relatively long period of subsequent development in this abdominal flap of tissue; and the young of the placental mammals are born at the most advanced stage of development of the three mammalian groups, but the absolute level of development varies from one species to another.

The emphasis in these chapters is on the human reproductive systems, but comparisons with other species have been included at various places in the text. The human species is *vertebrate*, more specifically it is included among the *mammals* and, even more specifically, humans are *primates* along with their many relatives among the monkeys, apes, and other species of a more exotic nature. Along with other vertebrates, we share many fundamental traits, but we are genetically closer to other mammals than to alternative groups of vertebrates such as reptiles or birds. Among the mammals, we are more like the gorilla, chimpanzee, and other primates than like different mammal groups to which horses, elephants, seals, whales, rats, mice, cattle, and other types belong. For some of the discussion, it may be important to know how we fit in the broad scheme of higher animal groups, and the few large categories mentioned above should be sufficient for our purposes.

The Male Reproductive System

The essential male reproductive functions are the production of the sperm (**spermatogenesis**) and the deposition of these cells in the female reproductive tract during copulation. The **testes** are the male sex organs, or gonads, that produce the sperm and also manufacture **testosterone**, the primary male sex hormone. Male sex hormones generally are known as **androgens**, and testosterone is the principal androgen manufactured by males.

Sperm are produced within the testes in the tissues of the **seminiferous tubules** (Fig. 5.1), and testosterone is manufactured in the **interstitial cells** of the sex organ. These two functions are interrelated since the processes of spermatogenesis require the presence and activity of adequate amounts of testosterone. Testosterone synthesis, on the other hand, is not dependent on sperm production. The androgen continues to be synthesized in the interstitial cells even if the seminiferous tubules are not functional. A lack of testosterone leads to sterility (no sperm produced) since this hormone is required for sperm production within the seminiferous tubules of the testes. Tying off the duct through which the sperm passes, whether for reasons of health or birth control, thus would have no effect on testosterone production; but sterility results because the sperm are blocked from their passage through the duct, even though sperm manufacture proceeds normally. Because testosterone synthesis continues normally (hence spermatogenesis, too), such a surgical sterilization procedure has no effect on male sexual behavior or the secondary sex characteristics, these being under hormonal control. There is essentially no sexual change that is effected other than the prevention of release of the sperm to the outside.

Castration does affect male sexuality since the testes are removed and therefore neither sperm nor testosterone production is possible. The

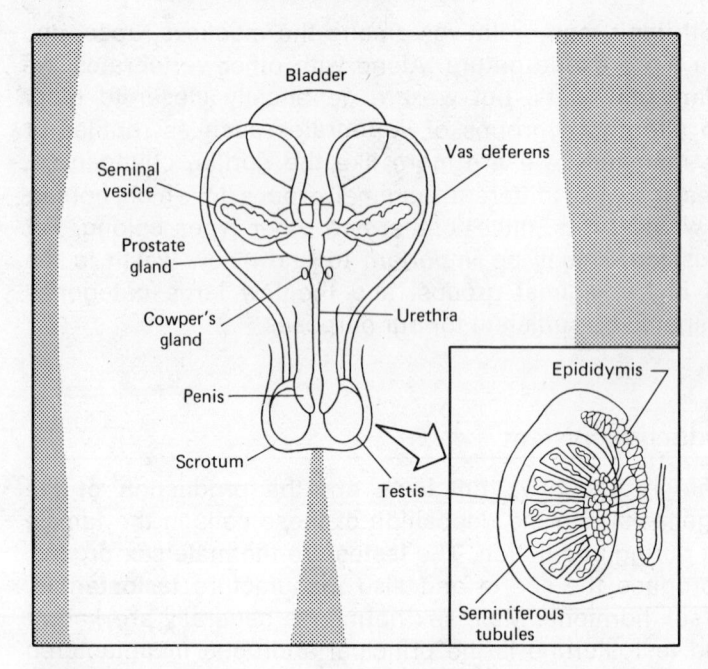

figure 5.1
The male reproductive tract, front view. Sperm produced in the seminiferous tubules of the testis move out through the epididymis to the vas deferens and then to the urethra. During sperm passage, seminal fluids are added from secretions of the seminal vesicles, prostate gland, and Cowper's glands. Semen and urine are released to the outside through the same urethral channel of the penis.

eunuch who guarded the harem or served as a palace functionary was a male who was castrated before puberty and whose secondary sexual traits never developed.

Reproductive Anatomy

The testes differentiate within the abdominal cavity of the male embryo and, in humans, they migrate to the **scrotum** shortly before or at birth. The scrotum is a wrinkled sac of skin that has little or no fatty insulation, unlike the rest of the abdominal area. In some mammals, such as whales and elephants, the testes remain within the abdominal cavity proper; in species such as seals and rats, the testes in the adult are closer in location to the groin area; and in bats and some rodents, the testes descend to the scrotum only during the breeding season.

If the testes fail to descend into the scrotum of the human infant, then the condition of **cryptorchidism** (*crypt*, hidden; *orchid*, testis) results. If this condition is not corrected surgically, then the testes atrophy soon after the boy reaches the age of puberty, and permanent sterility results. Each testis of the pair that is present in the adult human male is oval in shape and about one and one-half inches long. Fetal testes contain only undeveloped germ cells, but androgens are secreted by some of the hormone-producing cells so that a profound effect is exerted in the embryo leading to the development of male internal anatomy and external genitalia. The hormonal function of the testes continues uninterruptedly after birth and steadily accelerates near the time of puberty. The testicular secretions prepare the organ itself for effective reproduction and modulate the behavior and physical appearance of the individual male animal, including humans.

Sperm Formation

The testis is primarily composed of numerous, highly coiled seminiferous tubules. The peripheral layer of tubule cells consists of undifferentiated germ cells, called **spermatogonia**, which continually produce new germ cells by mitotic divisions (Fig. 5.2). The spermatogonia that migrate away from the tubule lining enlarge to become the primary **spermatocytes**. These are the cells in which meiosis takes place, each spermatocyte producing four immature sperm cells, or **spermatids**. Approximately 50 percent of the spermatids in humans carry the **X** chromosome, and the remaining half carry the **Y** chromosome; either of these occurs together with a set of 22 autosomal chromosomes.

Spermatids undergo a complex transformation to mature sperm, or **spermatozoa**, by mechanisms that are mostly unknown. The mature spermatozoan is differentiated into a *head* region and a *tail* (Fig. 5.3). The highly condensed nucleus is contained within the head, and a caplike structure partially covers the head. This covering structure is rich in enzymes that aid later in the digestion process that precedes the entry of the sperm into the egg at fertilization. The region of the tail that is adjacent to the head contains the energy-generating units of the cell, called mitochondria, and these are important to sperm viability and movement. The major structural basis for sperm movement resides in the whiplike portion of the tail, which may produce a speed of travel that takes the sperm a distance of 3 to 10 inches in one minute.

Each mature sperm takes an average of 60 to 72 days to be produced in humans. Although the whole process is lengthy, all stages of formation and development of sperm may be found in different portions of the seminiferous tubules or in different tubules within the testes. Animals that display seasonal breeding may undergo shrinkage of these tubules and

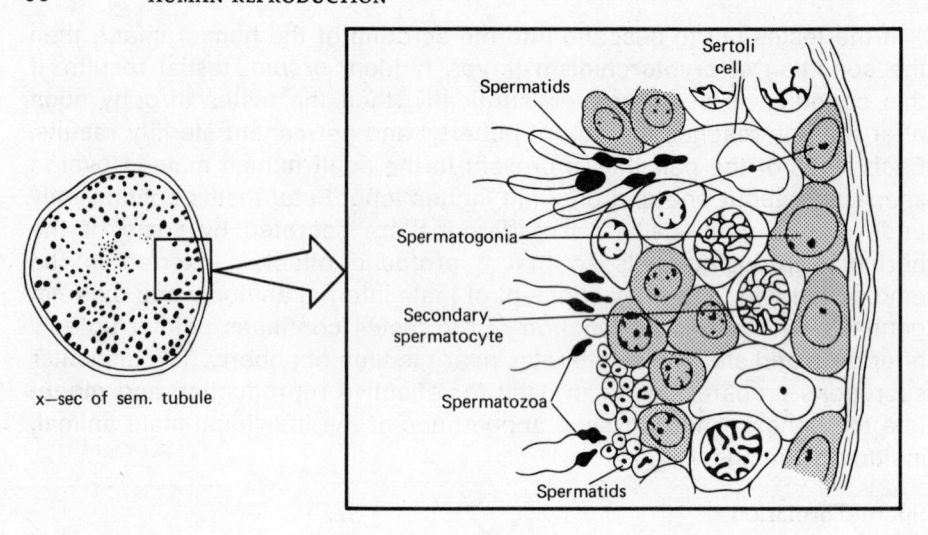

figure 5.2
Sperm formation occurs in the seminiferous tubules of the testis. Spermatogonia enlarge to become primary spermatocytes, which in turn undergo the first meiotic division to form secondary spermatocytes. Spermatids result upon completion of the second meiotic division, and these undergo a complex sequence of changes that lead to functional spermatozoa. Testosterone is secreted in the interstitial cells surrounding the seminiferous tubules.

degeneration of spermatogonia in the nonbreeding intervals. In humans and other continual breeders, there is no letup in the successive cycles of sperm production, and several hundred million sperm per day usually are produced by the adult human male.

Sperm production in humans proceeds normally only at a temperature that is a few degrees lower than the body's 98.6°F, for unknown reasons. Higher temperatures either inhibit sperm production or lead to sperm destruction; colder temperatures than the optimum only prevent spermatogenesis but do not destroy cells. The testes are pulled closer to the body at colder temperatures and are suspended from the body at higher temperatures by the contraction and relaxation, respectively, of the scrotal muscles.

Delivery of the Sperm

The sperm pass from the seminiferous tubules through a network of interconnected, highly coiled ducts that join to form the **epididymis**, and this in turn leads to the large and thick-walled **vas deferens**. This sperm passage system of the male reproductive tract performs several important

Human spermatozoon

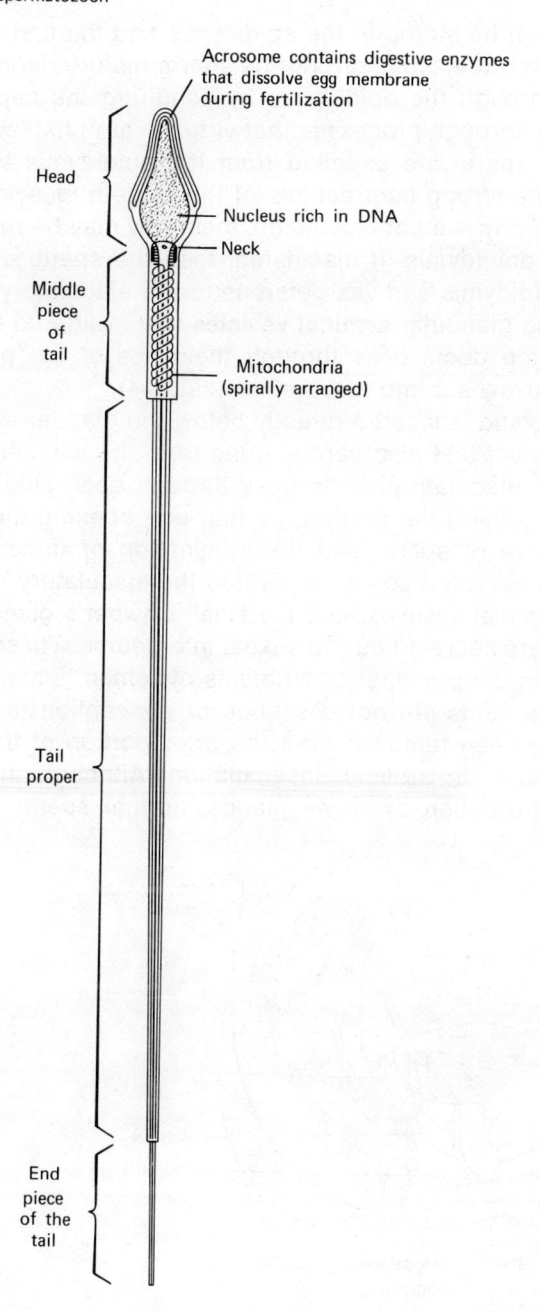

Acrosome contains digestive enzymes that dissolve egg membrane during fertilization

Head

Nucleus rich in DNA

Neck

Middle piece of tail

Mitochondria (spirally arranged)

Tail proper

End piece of the tail

figure 5.3
A mature human spermatozoan. The detailed structure is visible only with the electron microscope.

functions: (1) sperm can be stored in the epididymis and the first portion of the vas deferens prior to ejaculation; (2) the sperm mature during their storage or passage through the epididymis by acquiring the capacities for motility and fertility through processes that virtually are unknown; and (3) at ejaculation the sperm are expelled from the epididymis and the vas deferens due to the strong contractions of the smooth muscle lining the duct walls. If the sperm are not ejaculated, then they may be resorbed after a time from the epididymis. If ejaculated, then the sperm are propelled through the epididymis and vas deferens to the **ejaculatory ducts**, which include the large glandular **seminal vesicles** that drain into the vas deferens just before the ducts pass through the mass of the **prostate gland** to open by a narrow slit into the **urethra** (Fig. 5.4).

The large prostate gland is located directly below the bladder and surrounds the urethra, a duct that also carries urine from the bladder to the end of the penis. The ejaculatory ducts pass through each side of the prostate gland before joining the urethra, so that one opening functions in males for the release of sperm and the elimination of urine. Small quantities of fluids are secreted continuously into the ejaculatory duct by the prostate gland, seminal vesicles, and the small **Cowper's glands**; but much larger amounts are secreted during sexual intercourse. These fluids and the sperm cells are the principal constituents of **semen.** Experiments have shown that these fluids are not essential for conception to occur, since sperm that have been removed from the prior portion of the duct can be used successfully in artificial insemination. Although the fluid supply is a major contribution by these glands, normal sperm fertility

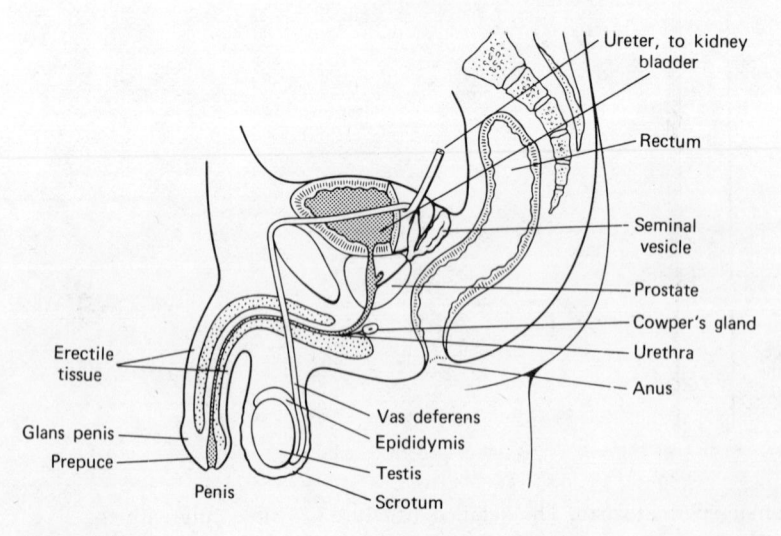

figure 5.4
The male reproductive tract and some surrounding structures, side view.

probably also depends on the energy-yielding nutrients in the fluid that is secreted by the seminal vesicles.

The recently discovered fatty-acid chemical derivatives, called *prostaglandins*, occur in especially high concentrations in human semen although they occur in various cell types, too. These chemical substances are secreted by the seminal vesicles and are added to the seminal fluids during ejaculation, probably aiding sperm movement within the female reproductive tract by inducing contractions of smooth muscles lining the uterus.

The small Cowper's glands that occur on either side of the urethra are responsible for the secretion of an alkaline fluid that is evident at the tip of the penis as a drop of clear, sticky material. This secretion appears prior to ejaculation and partially removes any urine that may be present in the urethra. It also functions in neutralizing the acidic environment of the female vaginal canal during intercourse, an essential condition for optimal sperm activity.

Erection

The male sexual act consists primarily of the **erection** of the penis, which permits entry into the female vagina, and **ejaculation** of the semen into the female reproductive tract (Fig. 5.5). The penis consists almost entirely

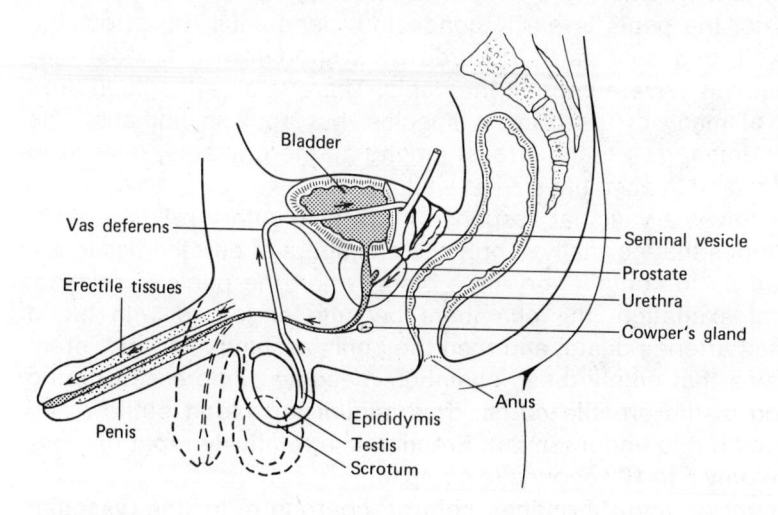

figure 5.5
The engorgement of blood vessels in the erectile tissues leads to penile erection. Ejaculation of semen occurs through the urethra, and the sperm are bathed in fluid secretions contributed from the various glands that line the urethra. The direction of sperm and semen movement is indicated by arrows. The pre-erection and post-erection positions of the penis and scrotal sac are shown in broken lines.

of three cylindrical cords of **erectile tissue** that essentially are spaces filled with blood vessels. The urethra is contained within one of these columns of tissue, the *corpus spongiosum*, and all three columns pass through the length of the penis. The corpus spongiosum enlarges at the tip of the penis to form the **glans penis**, a region that is rich in nerve receptors and thus quite sensitive to external stimulation. A free fold of skin (prepuce, or **foreskin**) overhangs the glans penis when it is relaxed. Circumcision involves the surgical removal of the foreskin, a ritual procedure in the Jewish and Islamic faiths.

Whereas all male mammals possess a penis, the males of most fishes, reptiles, and birds do not have a copulatory organ that can be inserted into the female reproductive tract. In most reptiles and birds, the **cloaca** serves as a passage for sperm and eggs and also for the elimination of body wastes. In species where the male lacks a penis, mating is accomplished by juxtaposition of the cloacas of male and female such that sperm are transferred to the female reproductive tract, and fertilization subsequently occurs within the female tract.

In most of the placental mammals, the penis is located in front, along the abdominal wall below or to the rear of the belly (for two-legged and four-legged postures, respectively). In many mammals, including horses and cattle, the penis is completely concealed within the preputial skin from which it emerges during erection. In humans, many other primates, and some bats, the penis is partly concealed, pendulous, free from the abdominal wall, and only partly covered by a short foreskin. Most carnivorous mammals have a bone in the penis. There is a cartilaginous rod in the penis of many of the primate species, but none in humans. The penis of the human male is the largest among all the primates, relative to adult body size and in absolute size as well.

Erection involves a vascular response to an involuntary reflex system. The small arteries that normally supply the chambers of erectile tissue are so constituted as to contain very little blood when the penis is relaxed. During sexual excitation, the chambers become engorged with blood since the small arteries dilate, and then the penis assumes its rigid, erect state. The veins that empty these chambers become compressed during the expansion of the erectile cords, thus minimizing blood outflow and further contributing to engorgement. Erection is a relatively rapid process that may take only 5 to 10 seconds to complete.

Nervous control and hormonal control contribute to the vascular changes that are evident in the erection phenomenon. Blood vessel engorgement occurs in response to stimulations and inhibitions of nerve cells in the circuits that affect the small arteries of the penis. Nervous stimulation of the Cowper's glands leads to their secretion of the alkaline mucous-like material that aids in neutralizing the acidic vaginal canal.

The primary input that initiates these reflexes comes from highly sensitive mechanoreceptors in the tip of the penis. But since the higher brain centers also exert control over nerve cell activity, thoughts and emotions can cause erection in the complete absence of mechanical stimulation of the penis. The failure of erection, or **impotence**, may result from organic causes as well as from psychological disturbances or effects of alcohol and other drugs on the higher brain centers.

There is no correlation between impotence and sterility, since a male may be fully capable of having an erection but may be sterile if he produces too few viable sperm or none at all. Impotent males may produce adequate quantities of functional sperm, but they are unable to inseminate the female since effective intromission requires a rigid penis.

Ejaculation

The process of ejaculation primarily involves a spinal reflex with the identical contributing nervous pathways that are involved in erection. When the level of stimulation reaches a critical peak, there is a patterned automatic sequence of nervous activity directed to both the smooth muscle of the genital ducts and the skeletal muscle at the base of the penis. Although the interactions are complex, the overall response occurs in two phases: (1) the contents of the genital ducts empty into the urethra as the ducts contract, and (2) the semen then is expelled from the penis (into the vagina) by a series of rapid muscle contractions. The sperm cannot enter the bladder nor can urine be passed during ejaculation because the opening at the base of the bladder is closed at this time. The rhythmical contractions of the penis during ejaculation are associated with intense pleasure during the entire event of **orgasm**. The entire body experiences a simultaneous and marked skeletal muscle contraction that is followed by the rapid onset of muscular and psychological relaxation. During orgasm there is a marked increase in blood pressure and heartbeat.

An average volume of three to five milliliters of the sticky, grayish-white semen may be expelled during ejaculation, containing about 100 million sperm per milliliter. There is considerable variation however, and it is generally believed that both the quantity and the quality of sperm constitute important factors for fertility. Older ideas had centered on quantity alone as the determinant for fertility. The orgasm subsides shortly after ejaculation, the penis becomes flaccid, and the male usually feels sexually satisfied.

Hormonal Control

All aspects of male reproductive function are influenced or controlled by the male sex hormones and the **gonadotrophic hormones** from the pituitary gland. The principal male sex hormone, or androgen, is the *steroid* testosterone that is secreted in testicular interstitial cells. The two gonadotrophic hormones are proteinlike substances that are synthesized in the anterior lobe of the pituitary gland: **follicle-stimulating hormone**, or **FSH**; and **luteinizing hormone**, or **LH**. These hormones were named originally on the basis of their effects in the female, but both FSH and LH are identical in molecular structure in the two sexes. The gonadotrophic hormones exert their effects only on testicular tissues in the male, whereas testosterone shows a broad spectrum of action: on testes, accessory reproductive structures, secondary sex characteristics, sexual behavior, and metabolism in general. Testosterone thus is directly influential in the development of both the primary and the secondary sex characteristics in the vertebrate male.

Effects of Testosterone

A deficiency of this hormone leads to sterility, since adequate amounts of the secretion are required for sperm production. Testosterone is secreted in the interstitial cells that lie scattered around between the seminiferous tubules. The hormone probably exerts its stimulatory effect locally, after diffusing from the interstitial cells into the surrounding seminiferous tubules where spermatogonia and spermatocytes occur. FSH and LH, along with testosterone, are required for spermatogenesis.

The function and structure of the entire male duct system, the lining glands, and the penis all depend on testosterone action. Upon castration in the adult, all of these accessory organs decrease in size, gland secretions are reduced markedly, and there is an inhibition of the smooth muscle activity of the ducts. Erection and ejaculation usually are deficient. All of these effects disappear upon treatment with exogenous testosterone.

A portion of the development and maintenance of sexual drive and behavior in the human male is testosterone dependent. Although castration may seriously impair these attributes, alternative patterns of sexual behavior appear to be completely unrelated to any excess or deficiency of this androgen. Various studies have shown that male homosexuals secrete testosterone in normal amounts at normal rates. Administration of exogenous testosterone may increase sexual drive, but sexual activity in these men remained homosexual. No clear correlation has been found to date between homosexuality or hypersexuality and hormonal status in either men or women, but controversial evidence does exist.

There are numerous manifestations of secondary sex traits in animals, all or most of which contribute in some measure to the success in gaining access to females. The stag uses its antlers in ritualistic fighting with other males during the breeding season; various courtship displays involve bright colors of plumage or skin; male vocalizations are an important aspect of reproductive behavior in frogs and toads, birds, and many other animals; and so forth. Virtually all of the obvious male secondary sex traits in humans are testosterone dependent; including the distribution of growing hair, lower voice pitch due to larynx growth, amount and distribution of muscle and fat, undeveloped breasts, and other characteristics. Boys who have been castrated before puberty develop no beard or axillary or pubic hair, and their voice remains higher pitched, among other traits. Male sopranos for some European church choirs sometimes were guaranteed by the practice of prepubertal castration.

Some of the effects of testosterone are described in terms of general body metabolism, but often these are difficult to separate from secondary sex characteristics. Body contours differ in males and females, more fat in women and more muscle in men leading to these effects. The increase in skeletal mass occurs under the influence of testosterone, along with a general stimulating effect on the overall growth of the individual. The common denominator of testosterone action is the promotion of growth, regardless of the region of the body in which its effects are exerted. But only certain types of cells respond to this androgen, and in ways that are qualitatively and quantitatively distinctive. The precise modes of action of this steroid hormone are unknown, but there undoubtedly are interactions with metabolic systems that are associated with the manufacture of proteins and nucleic acids.

Testosterone is not unique to the male. This hormone and other androgens occur in the blood of normal adult women and are secreted principally in the adrenal gland. Some androgen also is produced in the adrenal gland in males, but in small quantities and of a low potency. There is insufficient evidence, but it has been suggested that androgens may play some role in growth in the human female.

Neurohormonal Control of Testicular Function

FSH and LH are essential for normal spermatogenesis and for testosterone secretion; FSH stimulates sperm production directly, while LH stimulates the secretion of testosterone in the interstitial cells and thus indirectly affects spermatogenesis. If the anterior lobe of the pituitary gland is removed, the testes decrease in weight and there is a virtual halt to both spermatogenesis and the secretion of testosterone.

The **hypothalamus** is a region in the base of the brain, and it is attached by a stalk to the posterior lobe of the pituitary gland. The hypothalamus is associated with the anterior lobe through an unusual system of blood capillaries, but there are no direct nerve cell connections (Fig. 5.6). The capillaries provide a route for the transfer of relatively small molecules

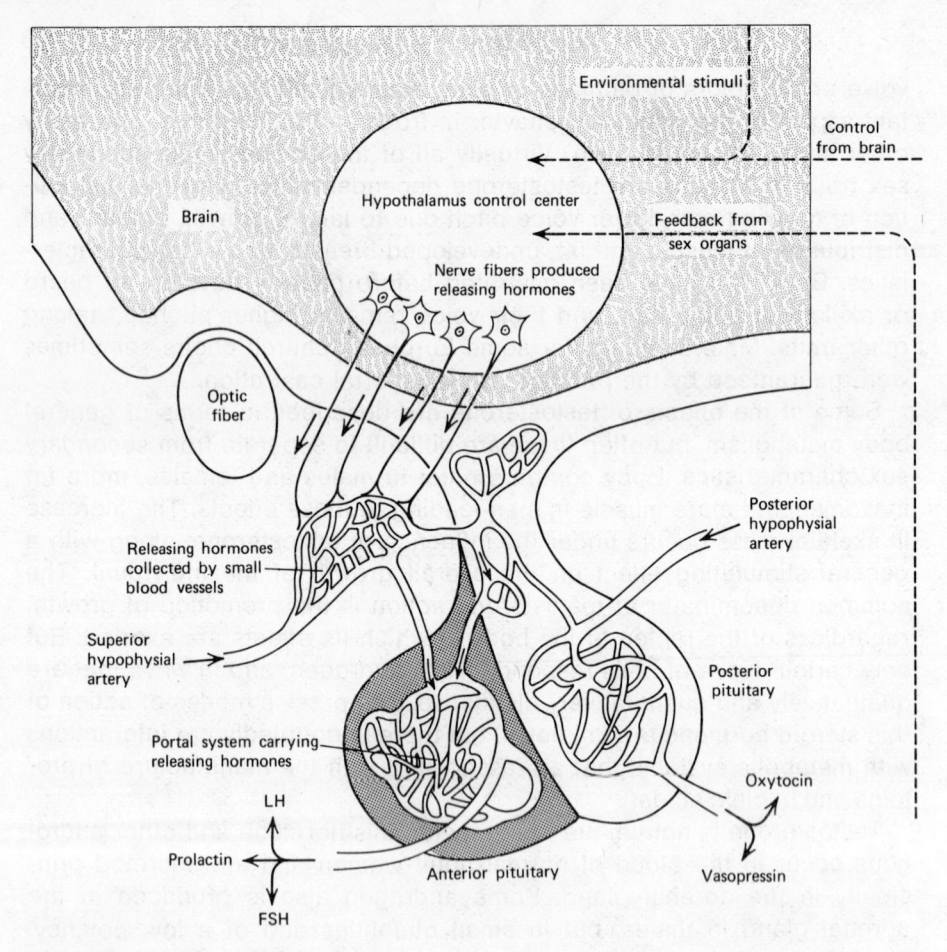

figure 5.6
Relationships of the hypothalamus and pituitary gland control system active in hormonal physiology. Releasing factors produced in the hypothalamus are transferred to the pituitary gland by an unusual blood portal system, and they stimulate the release of luteinizing hormone (LH) and follicle stimulating hormone (FSH) in both men and women. Another important hormone released from the anterior pituitary is prolactin, which governs milk production in mammals. Oxytocin and vasopressin are hormones that are produced directly in hypothalamic nerve cells and are stored in the posterior lobe of the pituitary gland until their release is stimulated, as occurs during the labor stages of delivery of the fetus and also during lactation.

called **releasing factors** from the neurons of the hypothalamus to target cells within the anterior pituitary. These releasing factors are hormones synthesized in brain cells, hence they are sometimes referred to as **neurohormones**. The neurohormones from the hypothalamus regulate the anterior pituitary secretion of yet other hormones, such as FSH, LH, thyroid-stimulating hormone, adrenal-stimulating hormone, somatotrophic (growth) hormone, and the hormone associated with milk production in nursing mothers, prolactin. The releasing factors are small molecules made up of amino acid units that are linked together. The releasing factor for LH recently was synthesized in the laboratory, and it is made up of 10 amino acids in tandem. Knowledge of molecular structures should prove useful in fertility—infertility therapy and in birth control measures (see Chapter 7).

There is a *feedback control mechanism* that operates at the level of the pituitary gland, very much like a thermostatically controlled furnace. As the temperature rises, the thermostatic control signals the furnace to shut off; subsequent cooling leads to a signal for heating to be resumed. Releasing factors similarly signal the anterior pituitary to secrete their complement of hormones. The LH-releasing factors move from the hypothalamus to target cells in the anterior pituitary, where the signal is received to turn on LH secretion. The LH from the pituitary travels to the testis where this gonadotrophic hormone stimulates the interstitial cells to secrete testosterone. But as the testosterone concentration increases, the LH action shuts down, thus reducing the level of testosterone secretion. The continued interactions among all these components of the system provide the basis for a maintained balance of hormonal output and testicular function.

There is a continuous, fixed rate of secretion of releasing factors in the nerve cells of the hypothalamus in the human male, so that FSH and LH in the anterior pituitary and testosterone in the testes are produced continuously and spermatogenesis occurs essentially at an unchanging rate from day to day. This situation in the male is very different from the cyclical swings of activity that characterize hormone production, gamete maturation and release, and reproductive capacities in the human female. Since the hypothalamus is a control center for reproductive physiology, it has become common to refer to a "male brain" and a "female brain" in relation to the continuous versus cyclical patterns of physiological activity. This set of terms is unfortunate in several ways, but especially if the functional activities of the hypothalamus are equated to the brain as a whole. One small part of this incredibly complex organ may be patterned differently between the sexes, and we don't really know that much about it, but that hardly merits the sweeping image of a "brain difference." Beyond this, the insistence on a "male brain" and a "female brain" has led to chauvinistic statements that might never have been made if the original distinction had been phrased more carefully and accurately.

The Female Reproductive System

As with males, there are two primary functions of the female gonad, or **ovary**: (1) the production of eggs, or **ova**; and (2) the secretion of the primary female sex hormones, **estrogens** and **progestogens**. The female receives the sperm from the male during coitus and, if fertilization occurs, then the fertilized egg enters a period of embryonic development during which growth and differentiation occur until the time of birth. The anatomical structures that are concerned with embryo development also function during the delivery of the baby, thus possessing a dual function in the female.

Unlike males, who continuously produce sperm and reproductive hormones in nonseasonal breeding species, females exhibit intermittent but cyclical patterns of gamete release and of modifications of structure and function in the reproductive system. In humans and other Old World primates these periodicities are manifestations of the **menstrual cycles**. Human males retain the capacity for sperm production and release from the time of puberty until some time in old age, which varies for different individuals. A recent newspaper story concerned an Iranian farmer, 130 years old, who had fathered 27 children by the several wives permitted under Islamic law, the youngest child being five years old. Women, on the other hand, produce and release mature ova for only a limited time after puberty. Ovum release generally stops at some time between the ages of 40 and 50, although at least two cases are recorded in which a woman of 57 gave birth to a baby.

All the immature germ cells of the female mammal are produced during the period of embryonic development, and no new ones are produced after birth. The human female fetus may contain as many as seven million **oocytes**, cells that have the potential to produce the ova, in the fifth month of pregnancy. This number declines afterward, but about one million normal oocytes are present in the newborn female. By the age of seven years, perhaps 300,000 of these normal cells still remain. This attrition has nothing to do with maturation and release of the ova, which does not begin until the time of puberty, generally between 11 and 15 years of age. The onset of puberty is signaled by the first menstrual discharge (**menarche**), which inaugurates the beginning of reproductive capacity. This capacity terminates at **menopause** when no more ova are released. Sexual activity continues to occur, however, under hormonal and other influences. During the 30 to 40 years of child-bearing age, only 350 to 450 ova from the original population of germ cells will be formed and released. Most of these cells degenerate, never having attained the stage of maturity.

There is considerable diversity among mammals in the age at which puberty is reached, but there is a general correlation such that larger animals take longer to attain sexual maturity than is true for smaller species. There is an absolutely longer interval between birth and puberty for the larger species of nonhuman primates as compared with smaller ones, but the human species reaches puberty the latest of all the primates. The period that extends from puberty to full physical maturity is known as **adolescence**, and it is synchronous with sexual maturity in some mammals. In others, there is a period of steady development of the sex organs and secondary sex characteristics during adolescence, before full sexual maturity is attained. Other modifications occur during adolescence, including growth. By and large, growth stops when full physical and sexual maturity is reached, although some bones in humans continue to elongate until the early or mid-twenties under the influence of anterior pituitary growth hormone secretion.

Reproductive Anatomy

The pair of ovaries occur in the body cavity in the lower abdomen, and the egg released from the ovary passes into the abdominal cavity and enters one of the two **oviducts**, known in humans as **Fallopian tubes**. These tubes empty into the pear-shaped **uterus**, where the embryo develops until the time of birth in all placental mammals. The uterus terminates in a muscular ring called the **cervix**, which projects a short distance into the single muscular tube of the **vagina** (Fig. 5.7). The vagina extends to the exterior and serves both to receive sperm during intercourse and as the birth canal when the baby is delivered. The external female genitalia, collectively called the **vulva**, include folds of fatty tissues known as **labia** and the **clitoris**, which is located at the junction of the two inner labial folds in front (Fig. 5.8). The clitoris develops from the same embryonic structure that forms the male penis, and it also is a sensitive erectile organ that responds to sexual stimulation. The opening of the **urethra**, which only serves a urinary function in females, is behind the clitoris; and the vaginal opening is just behind the urethra. The **hymen** is a thin membranous structure that partly occludes the vaginal opening and, unless destroyed by prior physical activity, it is ruptured during the first sexual intercourse. The function of the hymen is unknown.

From puberty until menopause, there is a succession of sexual cycles

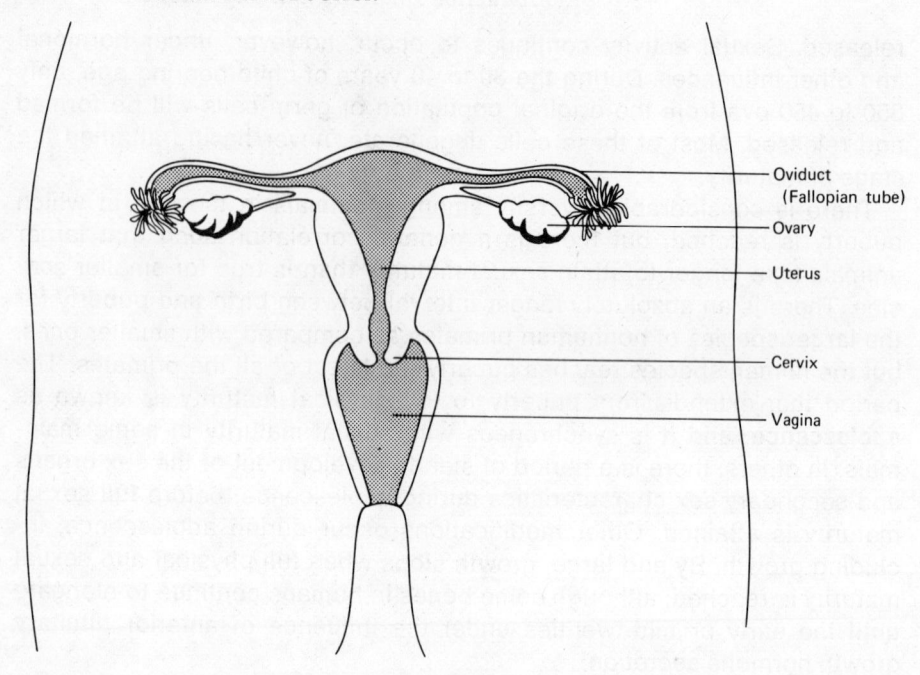

figure 5.7
The female reproductive tract, front view. Ova released from the ovary enter the Fallopian tube and proceed into the uterus, where inplantation in the uterine lining occurs if the egg has been fertilized. Sperm enter the uterus through the vagina and cervix, and these structures also serve as the birth canal when the fetus is expelled from the uterus at delivery.

exhibited by the female reproductive system, each lasting *an average* of 28 days. Considerable variation does exist in cycle length, but for simplicity we will use the average length. A period of **menstruation** marks the onset of each cycle during which some cells and blood are released through the cervix and vagina, lasting from 2 to 8 days, or about 5 days on the average. The biologically significant point in each cycle is the time of **ovulation**, or release of the ovum, which *generally* occurs on the thirteenth or fourteenth day after the start of the previous menstruation, but substantial variability occurs in this timing, too.

Growth of the Ovum and Follicle

Many cell clusters, known as primordial **follicles**, are present in the superficial layers of ovarian tissues. Each follicle is a cavity that consists of a single layer of flattened cells surrounding one immature oocyte. Of the million or so follicles that occur in the 2 ovaries of the newborn human

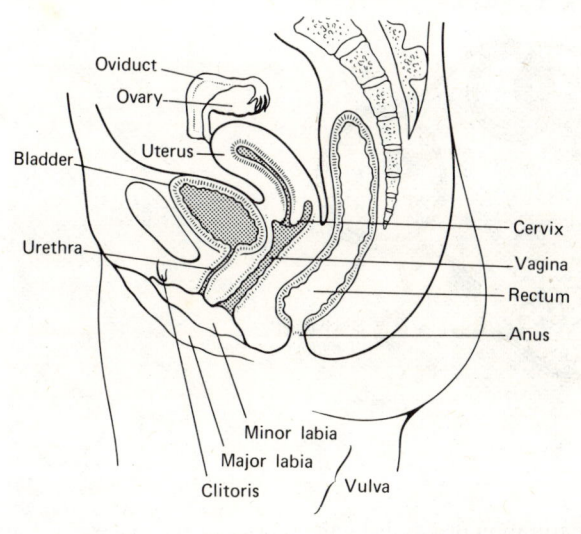

figure 5.8
The female reproductive tract and accessory structures, side view.

female, about 999 out of 1000 will not develop to maturity. The egg-producing germ cell, or **oocyte**, remains in an arrested early stage of meiosis from the time of its formation in the fetus until it degenerates or resumes meiosis and development in the fertile years of life. As the follicle develops (Fig. 5.9), on the average of one every 28 days in fertile women, there is a proliferation of surrounding cells and an enlargement of the oocyte within. Meiosis resumes to the completion of the first of the two nuclear divisions, but the second division does not occur unless the cell is penetrated by a sperm. This event induces the final stages of meiosis and the production of the functional egg nucleus, which then fuses with the nucleus brought in by the sperm. The result is the fertilized egg, which contains one nucleus that contains two sets of chromosomes, one set contributed by the egg and the other by the sperm nucleus that fused with it. Even though the final form of the female gamete is delayed until fertilization, it is usual to refer to the unfertilized cell as the ovum. Unlike the male spermatocytes that yield four spermatids from each meiotic cell, oocyte meiosis produces one functional egg and three abortive polar bodies at the conclusion of the division process. The bulk of the cytoplasm remains in the egg cell, and very little is lost in the small polar bodies. A rich supply of nutrients and cellular metabolic machinery thus is retained in the one functional gamete. Every human ovum contains one **X** chromosome plus 22 autosomal chromosomes.

The surface of the ovary bulges out as the follicle grows. Later, there is

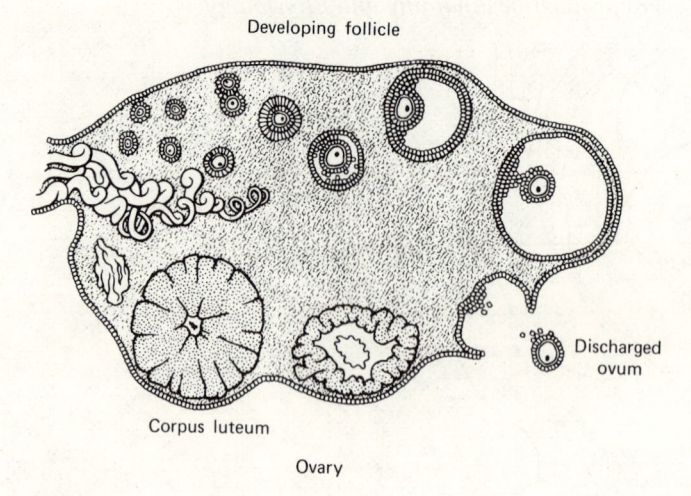

figure 5.9
The sequence of events during maturation of a follicle, the release of the ovum, and the formation of the corpus luteum from the remaining follicular materials. These stages occur each month during the operation of a menstrual cycle.

a rupture at this site through which the ovum is released, plus some adhering follicle cells, into the abdominal cavity to the Fallopian tube. Only one of the several developing follicles generally achieves maturity during a single menstrual cycle; the others degenerate for reasons that are unknown. Multiple births may occur when two or more eggs ripen at the same time and each is fertilized by a sperm (fraternal sibs), or if a single fertilized egg undergoes one or more divisions that lead to the development of two or more embryos (identical sibs). In some cases, there may be a combination of identical and fraternal siblings produced at a single birth.

Formation of the Corpus Luteum

After the oocyte of the ruptured follicle enters the Fallopian tube, the remaining cells of the follicle in the ovary enlarge and fill with fatty materials. This entire glandlike structure that develops in the ovary from the ruptured follicle is known as the **corpus luteum** ("yellow body"). The corpus luteum continues to develop from the inner glandular follicle cells for about 10 days but afterward disintegrates if fertilization has not occurred or if pregnancy has not been initiated. If the ovum is fertilized and embryonic development proceeds, then the corpus luteum continues to grow in the ovary and persists until near the end of pregnancy. If the corpus luteum breaks down because of lack of pregnancy, there is a degeneration of the endometrium that lines the uterus that had developed in anticipation of embryo implantation, and these uterine cells are included in the menstrual discharge.

The follicle cells release steroid sex hormones, collectively known as estrogens, during the **ovarian phase** of the sexual cycle (Fig. 5.10). These hormones contribute to the preparation of the endometrium for receiving the fertilized egg, along with secretions from other organs that lead to modifications of the uterine lining that are required for successful implantation of a fertilized egg. The corpus luteum is a different endocrine (hormone-producing) gland from the prior follicle form. Whereas the follicle cells principally produced the estrogen *estradiol*, the corpus luteum system begins to secrete quantities of the progestogen *progesterone* as well as estradiol. Both of these steroid hormones act to maintain the endometrium in an appropriate and receptive stage for successful implantation of the embryo during the **uterine phase** of the cycle.

Thus, this uterine phase is regulated by estrogen and progesterone that are manufactured in the follicle and its ovarian corpus luteum derivative. The earlier ovarian phase of the menstrual cycle, however, is controlled by gonadotrophic hormones from the anterior pituitary. The initial growth of the follicle apparently is independent of hormonal control, but further development depends on the presence of active follicle-stimulating hormone (**FSH**) and luteinizing hormone (**LH**). A negative feedback control system operates here as it did in the endocrine control of the male reproductive activity, as follows. A signal is sent to the hypothalamus that leads to FSH-releasing factor transport to target cells of the anterior pituitary, where FSH secretion occurs. The molecules of FSH travel from the anterior pituitary to the ovary where further development of the follicle occurs under the influence of this gonadotrophic hormone. But, as the estrogen level increases within the developing follicle, there is a feedback that shuts off further FSH secretion by the pituitary. The high estrogen levels also stimulate the LH-releasing factor from the hypothalamus, which in turn increases the secretion of LH by the anterior pituitary target cells. The LH from the pituitary acts in the ovary to stimulate the glandular cells in the cavity of the ruptured follicle to form the corpus luteum, hence the name "luteinizing hormone." If pregnancy does occur, then the high levels of estrogen and progesterone in the bloodstream shut off further manufacture of pituitary LH by this same feedback mechanism.

In summary, then, the changes that occur during the menstrual cycle are reflected first in the ovary and afterward in the uterus. The ovarian phase of the cycle is modulated by pituitary FSH and LH, under hypothalamus control, and the uterine phase that follows is influenced by the

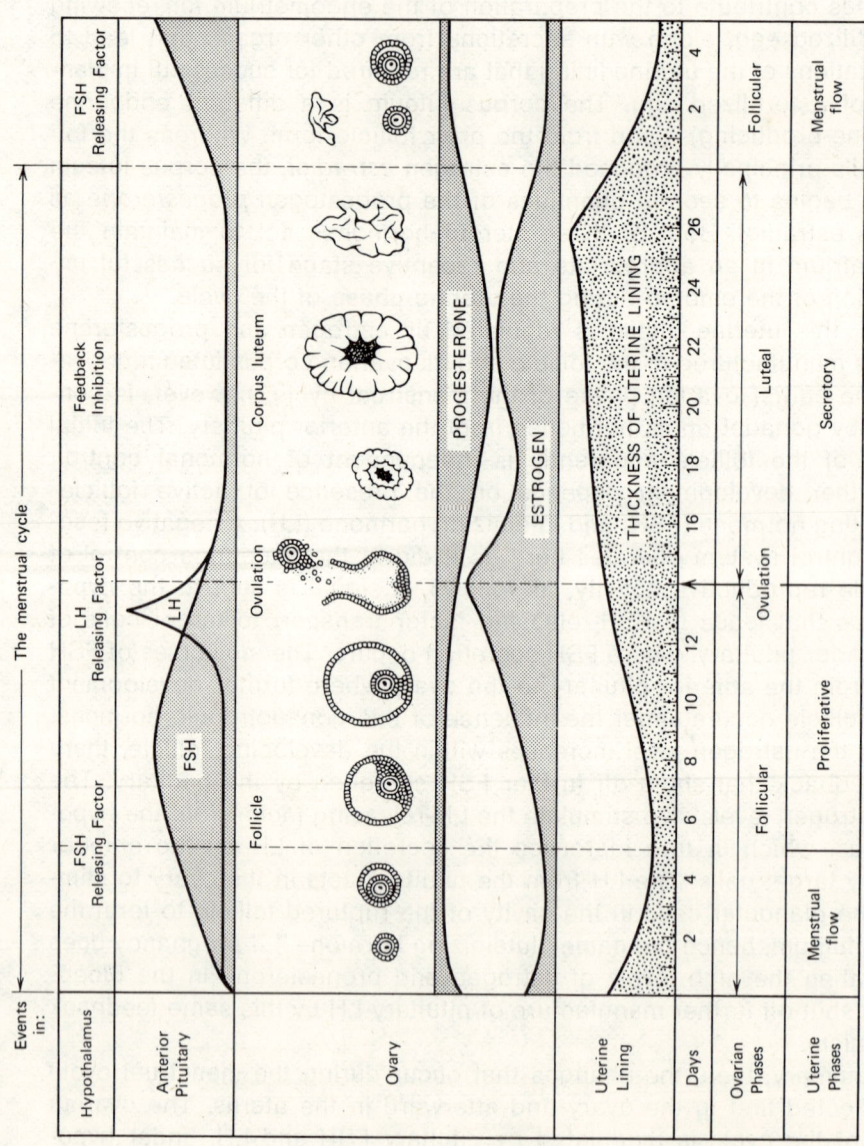

figure 5.10
The interrelationships between structures and hormonal physiology during the menstrual cycle in the human female.

ovarian steroid sex hormones estrogen and progesterone. The changing balances of pituitary and ovarian hormones, operating under negative feedback control, thus characterize the sexual or menstrual cycle (Fig. 5.11). While the corpus luteum is present, an enormous amount of estrogen and progesterone is manufactured, inhibiting further FSH and LH secretion. When the corpus luteum degenerates in the absence of pregnancy, the levels of FSH and LH rise again as the concentration of ovarian hormones in the bloodstream decreases. With this next rise in the amount of pituitary gonadotrophic hormones, a new follicle is stimulated to mature and a new menstrual cycle is initiated. Except during pregnancy, these events recur every 28 days, on the average, between the ages of puberty and menopause.

The oral contraceptive "pill" controls fertility by inhibiting ovulation. The synthetic estrogenlike and progesteronelike hormones in the contraceptive are introduced at a sufficiently high concentration to cause the shutdown of pituitary gonadotrophic secretion, hence producing a failure of follicle maturation and ovum release. Daily use of the pill for 21 days of an established 28-day menstrual cycle leads to a thickening of the uterine lining; but when the woman stops taking the pill for the next

	Follicular Phase	Ovulation	Luteal Phase	
Hypothalamus	FSH Releasing Factor	FSH Releasing Factor	LH Releasing Factor	FSH Releasing Factor
	↓	↓	↓	↓
	Anterior pituitary	Anterior pituitary	Anterior pituitary	
	FSH	FSH	LH	
Ovary	Follicle	Follicle	Corpus luteum	
		Estrogen	Progesterone and estrogen	
Uterus		Uterine lining	Uterine lining	

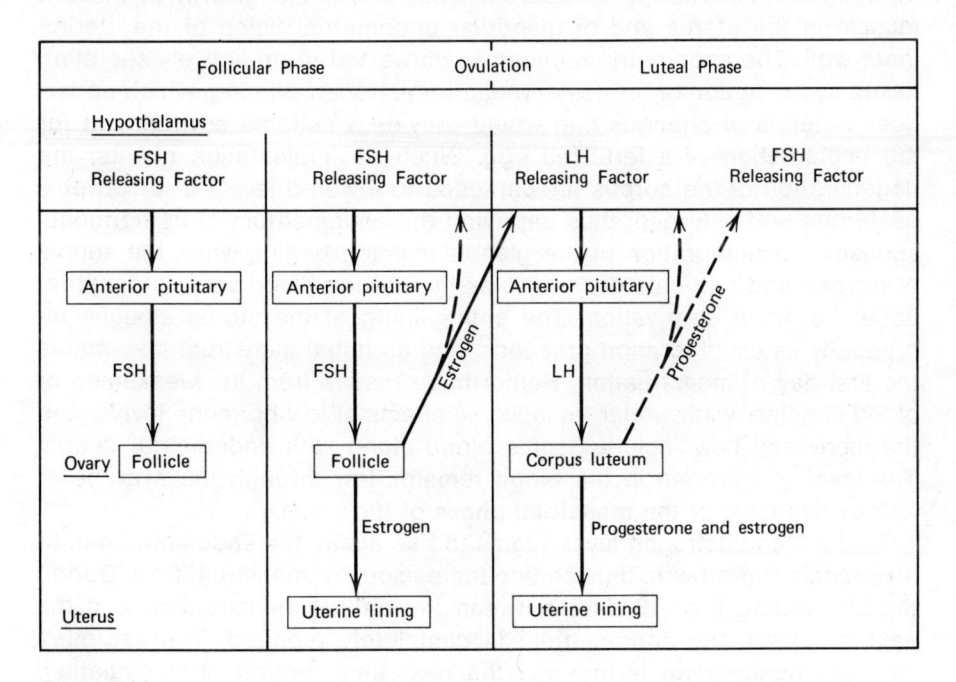

figure 5.11
The feedback control system that regulates the changing balance of hormonal production and effect during the menstrual cycle.

7 days, then the endometrium is sloughed off and a menstrual period is produced even though ovulation has not occurred.

In animals that ovulate only after copulation, nervous impulses are transmitted from the stimulated female reproductive tract directly to the hypothalamus, which in turn stimulates the secretion of releasing factors that induce anterior pituitary manufacture and the release of FSH and LH. Ovulation in humans and other primates occurs without relation to sexual intercourse, but the mechanisms that lead to the cyclical bursts of gonadotrophic hormones and subsequent ovulation largely remain unknown. There is some recent evidence, however, which indicates that some women may ovulate in response to intercourse. This would have important consequences for couples who use the rhythm method to achieve birth control, since induced ovulation would not conform to any program of periodic abstinence from sexual intercourse. Further evidence is required to substantiate the preliminary information that is available.

Uterine Changes and Menstruation

Estrogen and progesterone influence profound changes in uterine tissues during the menstrual cycle. Estrogen stimulates the growth of smooth muscle in the uterus and of glandular endometrial lining of the uterine inner wall. The endometrium becomes converted to an actively secreting tissue by the action of progesterone, after estrogen priming, which undergoes a series of changes that would provide a suitable environment for the **implantation** of a fertilized egg. When no implantation occurs, the degeneration of the corpus luteum leads to lowered levels of blood progesterone and estrogen, thus depriving the endometrium of its hormonal sponsors. Disintegration of the glandular lining begins when the supply of oxygen and nutrients decreases as the uterine blood vessels constrict under hormone deprivation. The entire lining of the uterus sloughs off gradually as disintegration proceeds, and an initial menstrual flow marks the first day of menstruation. Hemorrhage results from the weakening of blood capillary walls under the adverse effects of low hormone levels, and the menstrual flow includes some blood along with endometrial debris. The level of estrogen in the blood remains low through the average of four or five days of the **menstrual phase** of the cycle.

As the blood estrogen level begins to rise again, the endometrium initiates repair and growth, thus ending the episode of menstrual flow. During the intervening 9 or 10 days between the end of menstruation and the next ovulation, the endometrium is completely repaired. The secretory type of endometrium is induced the next time around after ovulation, when the corpus luteum forms and ovarian hormone concentrations in the blood increase. The cycle is completed when the corpus luteum degenerates. Overt menstruation occurs only in Old World monkeys, apes, and humans, although slight bleeding from the uterus occurs in many other primates upon degeneration of the endometrium.

General Effects of Female Sex Hormones

Estrogen is primarily responsible for the maintenance and development of the entire female reproductive tract (uterus, oviducts, and vagina), the glands lining this tract, the external genitalia, the breasts, and the pelvis. The spectrum of secondary sex characteristics also includes differential growth of the hair follicles distributed over the body, and the amounts and distribution of fat and skeletal muscle mass in the body. The effects on ovarian and uterine tissues already have been mentioned. In addition to all of the secondary sex paraphernalia, the differentiation of the embryonic gonad to a functioning ovary, which is the primary sex characteristic, also is under the control of estrogen.

The effects of progesterone are less widespread in place and time; the principal site of action of this steroid hormone being the endometrium during the luteal phase of the menstrual cycle. Progesterone has significant effects on the breasts and the uterine smooth muscle during pregnancy. The slight elevation in body temperature that occurs in many women about 10 days after menstruation ends is attributable to progesterone, but the mechanism that produces this effect is unknown. This increase in body temperature may serve as a crude indication that ovulation had occurred earlier, since the rise in temperature accompanies the formation of the corpus luteum under progesterone stimulation.

Estrogen does not influence growth as substantially as we know to be the case for testosterone. It is believed that the spurt of body growth that occurs at puberty is due to high levels of blood estrogen in females at this time, but the hormonal effects are generally less pronounced than those of testosterone in boys at the age of puberty. Estrogen also occurs in low concentrations in males, synthesized in the adrenal gland and in the interstitial cells of the testis.

The Estruos Cycle and Fertility

Except for humans, all mammals are reproductively active only during specific breeding times or seasons of the year. The time of reproductive activity coincides with the period of fertility and is known as **estrus**. Most mammals have one or more breeding seasons per year, which may be exhibited as a long sustained estrus or a polyestrous situation in which a succession of sexual cycles may occur (Table 5.1). Regardless of the duration or frequency of estrus, ovulation always occurs at a time of

table 5.1
Reproductive Patterns in Some Mammals

Species	Type of Cycle	Length of Estrus	Ovulation	Menstruation
Rabbit	monestrous	continuous, May–July	induced	none
Dog	monestrous	7–9 days, spring and fall	spontaneous	none
Cat	monestrous	9–10 days, spring and fall	induced	none
Mouse	polyestrous all year	4–6 days	2–3 hours after start of estrus	none
Cow	polyestrous all year	17–23 days	10 hours after end of estrus	none
Horse	polyestrous March–Oct.	21–23 days	toward end of estrus	none
Chacma baboon	polyestrous all year	29–42 days	spontaneous	4–9 days
Chimpanzee	polyestrous all year	34–35 days	16th day onwards	2–3 days ave.
Human	anestrous; menstrual	21–34 days, 28 days ave.*	6th to 23rd day, 14th day ave. *before menstruation*	2–8 days; 5 days ave.

* Figures given for menstrual cycle; there is no estrus period in humans. (From R. J. Harrison and W. Montagna, *Man*, 1st ed., pp. 235–238. © 1969, Meredith Corp. Used with permission of Appleton-Century-Crofts.)

maximum receptivity of the female during estrus. Ovulation may be induced by the stimulus of copulation in mammals that have a single estrus, but usually it is spontaneous. The length of a sexual cycle varies from about 4 days (rats and mice) to 35 days (chimpanzees) in mammals that are polyestrous. Human females are unique, as far as we know, in lacking an estrous pattern of reproductive activity. Women are sexually receptive throughout the year, during infertile periods as well as at the time of ovulation every 28 days; and ovulation occurs uninterruptedly in successive cycles in nonpregnant women. Furthermore, sexual activity does not cease after menopause even though fertility has terminated. The relationship between sexual drive and hormonal status is unclear, but it is known that female sex drive is not altered by removal of the ovaries or by menopause, after which ovarian function ceases for physiological reasons.

Primates of the groups of Old World monkeys and the great apes display striking and conspicuous changes in the sex skin around the genital region in midcycle. The sex skin may become highly pigmented as in the macaques, markedly swollen as in chimpanzees, and also may occur elsewhere on the body as seen in gelada baboons, which have a sex skin on the chest as well as around the genitalia. Among some of the subhuman primates, the genitalia and the sex skin may become swollen at the time of maximum receptivity in the female. In addition to these traits, many other factors may indicate female receptivity to the prospective mate. Chief among these are olfactory signals due to secretions from specialized glands in the genital and anal regions, as well as other body areas in different mammalian species.

Development of Biological Sex

Until about the sixth week of embryonic life, the sex of the developing system is undetermined anatomically. Every embryo contains one pair of primordial gonads each with the potential for developing either as an ovary or as a testis; two sets of ducts, one of which can differentiate into the male internal reproductive tract and the other set that has the potential for development into the female reproductive structures; and ambiguous external genitalia that are indistinguishable as male or female. The first of these systems to differentiate is the primary sex trait of the ovary or testis from the indifferent gonad. The specification of the gonad begins in male embryos at about the sixth week, but gonad differentiation as an ovary is not begun in the female until about the twelfth week, according to recent evidence.

Under normal conditions, the primordial gonad differentiates as a testis if a **Y** chromosome is present, but it differentiates as an ovary if there are at least two **X** chromosomes and no **Y**. The core tissues of the sex organ undergo testicular differentiation in males, whereas ovarian development proceeds from the outer layers of gonadal cells in the female fetus. Once the gonad has differentiated as an ovary or testis, there is no reversal in mammal species, although gonad functional reversal can occur in lower vertebrate forms.

Once the gonads have been determined, under the influence of the sex chromosomes, then the internal reproductive tract differentiates under the influence of hormonal secretions while the external genitalia remain ambiguous and undetermined (Fig. 5.12). The androgens produced by the testes of the embryo influence the primordial male ducts to form the vas deferens, seminal vesicles, and the ejaculatory ducts. The primordial female ducts become reduced to vestigial remnants in the male fetus, presumably under the influence of some inhibiting chemical substance for which there is indirect evidence. If an ovary has been determined, then the primordial male ducts become reduced to remnants while the primordial female ducts differentiate to form the uterus, the Fallopian tubes (oviducts), and the upper portion of the vagina. If the gonads are removed or if they fail to develop for any reason, then the primordial female ducts still will differentiate to form the internal female reproductive structures. Female internal structures thus will differentiate in either the presence or absence of ovaries, so long as there are no androgens to direct male structural differentiation. Embryos that are **XY** can differentiate internal female structures if the testes fail to secrete androgens or if the body cells are unresponsive to androgens, as in some clinical conditions (see Chapter 11). Androgens are required to direct differentiation toward male and away from female structures, but these steroid hormones must be present at particular and critical times during development.

The final step in the development of sexual anatomy involves the differentiation of the external genitalia in the fetus (Fig. 5.13), under control of sex hormones. The undetermined external structures consist of a **genital tubercle** above a **urogenital groove**, around which occur the **urethral folds**, and the **labioscrotal swellings** that are next adjacent to these folds. The changes that take place during the third month of fetal life lead to functionally and structurally distinct external genitalia that develop from structures that are homologous in males and females. In the male fetus, the genital tubercle tissues become incorporated into the penis, which becomes enlarged; the urethral tube becomes enclosed within the penis when the urethral folds fuse; and the scrotum forms when the labioscrotal swellings fuse at the midline. The tissues of the genital tubercle in the female fetus form parts of the clitoris, which develops from a slightly enlarged tubercle structure; the urethral folds remain separate and differ-

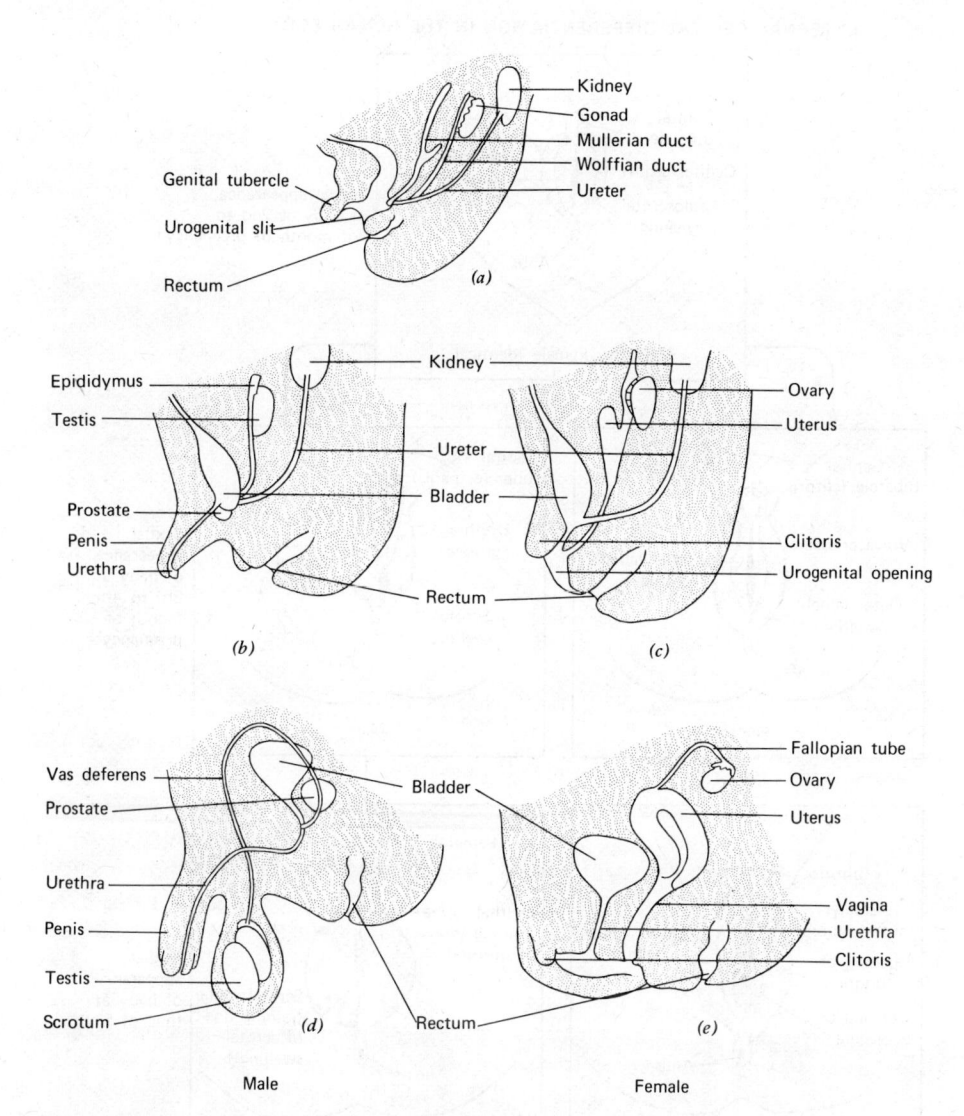

figure 5.12

The development of male and female reproductive structures during intrauterine life. (a) Sexually undetermined embryo, approximately in the sixth week of development; (b) male fetus in the third month of development; and (c) a female fetus of approximately the same age. (d) Male and (e) female structures as they are developed at the time of birth of the infant.

EXTERNAL GENITAL DIFFERENTIATION IN THE HUMAN FETUS

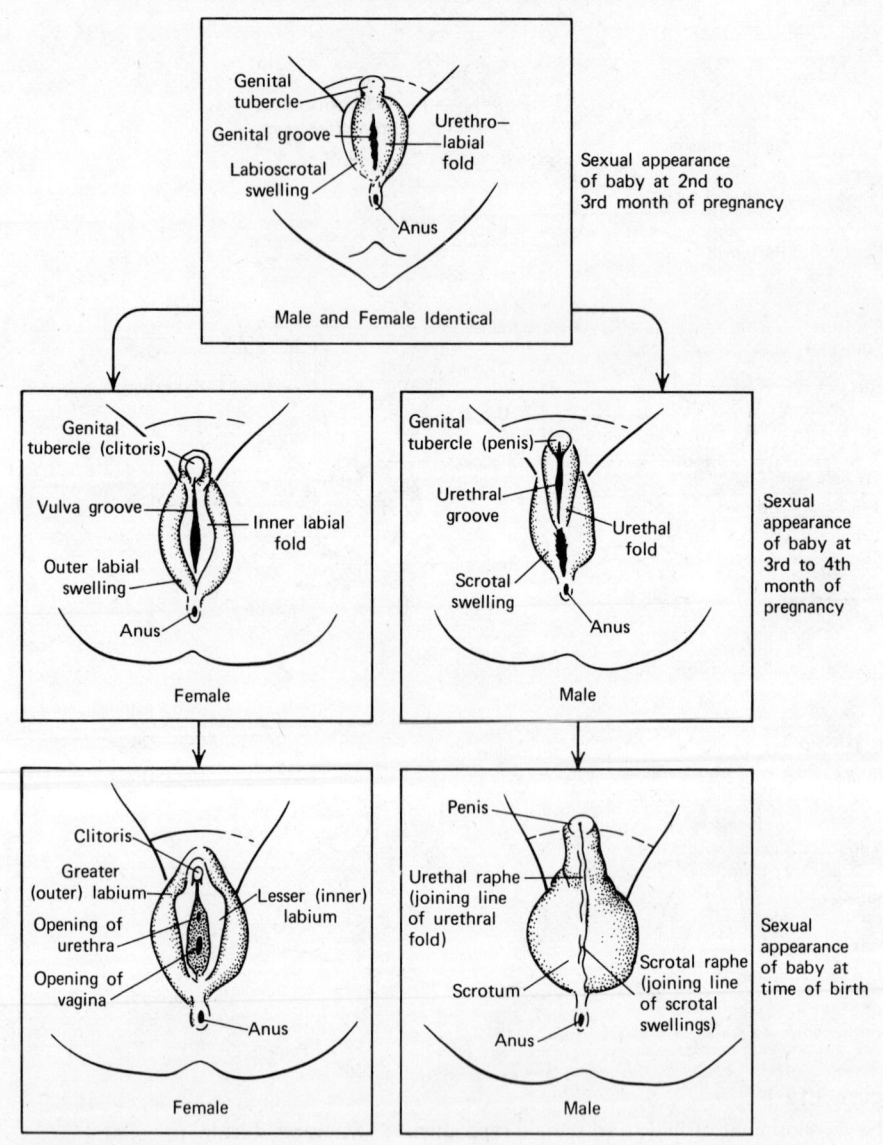

figure 5.13

Appearance of external genitalia in the undetermined stage (beginning of the third month) and their differentiation into distinct male and female structures (approximately at the end of the fourth month) from rudiments that are homologous in origin.

entiate to become the minor labia while the major labia form from separate labioscrotal swellings; and portions of the vagina differentiate from the urogenital groove, posterior to the urethral opening.

Suggested Readings

Bonner, J. T., Hormones in social amoebae and mammals. *Scientific American, 220* (June 1969), 78.

Guillemin, R., and R. Burgus. The hormones of the hypothalamus. *Scientific American, 227* (November 1972), 24.

Harrison, R. J., and W. Montagna, *Man.* New York: Appleton-Century Crofts, 1969.

Jost, A., A new look at the mechanisms controlling sex differentiation in mammals. *Johns Hopkins Medical Journal, 130* (1972), 38.

Katchadourian, H. A., and D. T. Lunde. *Fundamentals of Human Sexuality.* New York: Holt, Rinehart, Winston, 1972.

McCary, J. L., *Human Sexuality.* New York: Van Nostrand Reinhold, 1973.

Perry, J. S., *The Ovarian Cycle of Mammals.* New York: Hafner Publishing Company, 1972.

Zuckerman, Sir Solly, Hormones. *Scientific American, 196* (March 1957), 76.

chapter six

Fertilization, Pregnancy, Birth, and Lactation

The relative success of sexual reproduction in mammals depends on many sets of factors and conditions. Among these are the relatively short lifetime expectancies of the eggs and sperm themselves. In humans the sperm usually remain functional for about 48 hours after ejaculation into the vagina, although an ovum may retain fertility for only 10 to 15 hours, to cite some average figures. There is considerable variation within our own species, as well as among different mammals. The sperm of the stallion may retain fertility for up to five days within the reproductive tract of the mare, and bats have been reported to be able to store sperm in the female tract for five or six months, so that newly shed ova may be fertilized by sperm deposited before hibernation. Bull semen can be stored for years at extremely low temperatures ($-80°C$ or $-185°F$) and provide active sperm for artificial insemination of domestic cattle. Human sperm has been stored at 4° C (refrigerator temperature) and has retained fertility under controlled laboratory conditions. These observations have prompted suggestions for establishing sperm storage banks from which viable gametes may be obtained long after their release in an ejaculate. Others have further suggested that sperm banks might provide materials for controlled breeding in population improvement programs in the future, which some people have shudderingly compared with Huxley's *Brave New World*.

Although the gametes may survive for longer than the average number of days or hours, ova and sperm that have not aged too much since release from the gonads seem to have the greater probabilities for success. Viable pregnancies are much more likely to occur in those mammals that ovulate in response to the stimulus of copulation. In such induced ovulators the preponderance of the sex cells will be of optimum age, and there is high certainty that an ovum will be available for the sperm that are introduced by the copulating male. For humans and other species that do not ovulate regularly by such a reflex system, there is a higher rate of pregnancy failure for various reasons, including the absolute age of the available gametes at the time of coitus.

Fertilization and Implantation

Once released from either one of the two ovaries into the abdominal cavity, the ovum enters the adjacent Fallopian tube (Fig. 6.1). The oocyte, which will become the functional egg, is sucked into the tube by the actions of the beating, hairlike cilia that line the fingerlike projections at the end of the oviduct, and by the smooth muscle contractions that are instituted in these regions at ovulation. On rare occasions, the ovum may remain in the abdominal cavity and perhaps be fertilized by a sperm that reaches it through the open end of the oviduct. If the pregnancy proceeds normally, then the baby may be delivered by a cesarean section (delivery through the opened abdominal cavity).

The ovum moves rapidly down the oviduct for several minutes, but very soon it slows down because the contractions of the smooth muscle of the passageway diminish. It may take several days for the ovum to reach the uterus, so that fertilization must occur within the oviduct since the unfertilized egg has a very brief life span. The first sperm may arrive at the site of fertilization within the Fallopian tube about fifteen minutes after deposition in the vaginal canal, so that sperm movement probably is aided substantially by external factors and not by its own efforts alone. Once in the uterus, the sperm are moved toward the ovum principally by upward contractions of the smooth muscles of the uterus and Fallopian tube. Sperm movement by its whiplike tail is important in the final stages of approach and penetration of the ovum. Recently discovered fatty-acid substances (*prostaglandins*) are responsible for some of the muscular contraction episodes that are involved in fertilization, as well as many phenomena not associated with reproduction. Regardless of the details and of such enigmatic features as the mechanism by which the ovum travels down the oviduct while this tube simultaneously aids in directing the sperm upward, only about 100 to 1000 of the several hundred million

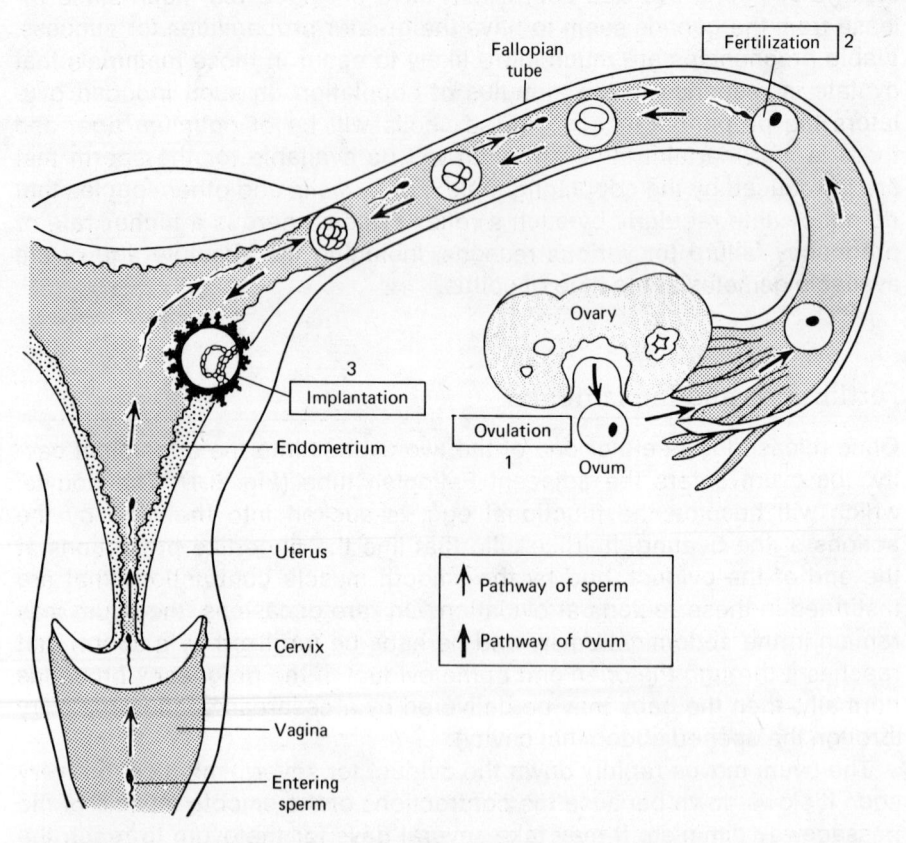

figure 6.1
Events during the week between fertilization of the ovum and the implantation of the embryo-containing blastocyst in the endometrium lining the uterus.

sperm actually reach the oviduct after the ejaculate has been deposited in the vagina. Most of this attrition is due to scavenging white blood cells that react to any foreign substances, and it occurs mainly in the cervix. This observation provides one basis for understanding the requirement for a minimum number of viable sperm to be present in the ejaculate if fertilization is to be effected. It has been estimated that at least 35 million functional sperm (about 25 percent of the average human ejaculate content) must be present at one time if the man is to be fertile. Infertility results from any one or more causes, including the presence of an excess of abnormal sperm even if the total count seems adequate to sponsor a pregnancy.

Fusion of the Egg and Sperm

There is no particular evidence for the presence of attractant chemicals, and it is generally assumed that the surviving sperm reach the egg by random motion. Many of these survivors reach the ovum, but only one sperm successfully initiates pregnancy by penetrating and fertilizing the egg cell. Although we know very little about the physiology of mammalian fertilization, there is some cross-specificity such that only sperm can penetrate the ovum, and the ovum in turn is the only cell that serves as the target for the sperm. The spermatozoan that eventually enters the egg reaches the interior after dissolving its way through the surrounding follicle cells and one site of the egg membrane itself. The enzymes in the caplike structure covering the head of the sperm (see Fig. 5.3) undoubtedly expedite these activities. A fertilization membrane forms almost immediately around the penetrated egg, and other sperm are prevented from entering. By mechanisms that are largely unknown, the fertilized egg remains viable only if a single sperm has gained entry. In humans and other mammals, the penetrated ovum itself degenerates if more than one sperm has entered, or the excess sperm die before reaching the cytoplasm in the relative vastness of the egg cell.

Although we have referred to the ovum throughout these discussions, the cell that is released from the ovary is still an oocyte that only has reached the end of the first of the two meiotic divisions when it enters the oviduct. Upon penetration by the sperm, the oocyte nucleus rapidly completes the second meiotic division and develops into a true ovum. Only one of the resulting meiotic nuclei is retained in the egg cell, and this nucleus fuses with the nucleus of the entering sperm to produce the first of the nuclei of the new-generation individual. During this process, the sperm loses its tail and its nucleus enlarges from the greatly compacted state that was typical of the spermatozoan. The fertilized egg contains an equivalent genetic program from the two gametes, but it derives most of its structural and chemical machinery from the contents of the egg cell proper.

If it has been fertilized, the egg continues its travels down the oviduct to the uterus, undergoing developmental changes during this time period. If the fertilized egg remains in the Fallopian tube, as occurs occasionally, implantation may take place, but the tubal pregnancy is unsuccessful because there is not enough space for the embryo or fetus to grow. Surgery may be necessary if the growing embryo causes rupture of the oviduct and consequent hemorrhage, otherwise the woman may die. If the egg is not fertilized, then the ovum disintegrates and is reabsorbed into the surrounding tissues.

Implantation

The fertilized egg undergoes a series of cell divisions during the two- to three-day journey from the oviduct to the uterus (see Fig. 6.1). The one cell divides to become two; then these divide to four-celled, eight-celled, and higher numbers until a solid cluster is formed just prior to its entry into the uterus. At this time the solid mass becomes transformed into a hollow ball of cells known as a **blastocyst**. The blastocyst consists of a little more than 100 cells at this stage and measures about 1/250th of one inch in diameter. Within the fluid-filled sphere that is one cell layer thick at its periphery, the primordial embryonic cell mass is attached at one site.

The blastocyst floats free in the uterine fluid for several days and continues to increase in mass by mitosis and cell divisions. While the embryo develops between day-14 (fertilization of the ovum) and day-21 of the menstrual cycle in the human female, the uterine lining simultaneously undergoes preparations for **implantation** of the embryo in the presence of estrogen and progesterone secretions (see Fig. 5.10). The embryo begins implantation approximately six or seven days after fertilization, which is approximately the twentieth or twenty-first day of the menstrual cycle. The blastocyst adheres to the endometrial lining way up on the back wall of the uterus, and a series of profound changes characterize the blastocyst itself and the endometrium in which it has become implanted within the uterus. The outer cell layer of the blastocyst enlarges rapidly and produces digestive enzymes that break down endometrial tissue and pave the way for the embedding of the embryo cell mass that is still within the blastocyst. In about three percent of human pregnancies, the blastocyst becomes implanted in some place other than the endometrial lining of the uterus. These *ectopic* pregnancies may occur in the ovary, abdominal cavity, or more commonly in the Fallopian tube, as mentioned earlier.

There is no evidence of a delay in the implantation of the human blastocyst once it reaches the uterus. Such a delay occurs characteristically in some mammalian species in relation to lactation, such a longer interval separating in time the arrival of a new group of offspring while others from a previous pregnancy still are suckling. Still other factors may lead to a delay of implantation, unrelated to lactation, perhaps for several months as an average for various mammals. Subhuman primates, like humans, show no evidence of implantation delay.

Once it is implanted, about five days after reaching the endometrium, the developing embryo begins to receive nutrients and other materials by absorption from the endometrium and the endometrial debris that was liberated after digestive enzyme action by the embedding embryo. This arrangement is adequate only for a short time and, by the age of two or three weeks, the embryo begins to receive maternal nutrients and oxygen through a system of blood vessels that has differentiated. Still further modifications occur in cells and tissues until the **placenta** has developed and an adequate embryonic circulation has been established. The embryo is usually called a **fetus** after the end of the second month following conception, although it remains an embryo biologically until birth.

The Placenta

The placenta is a uniquely mammalian organ that is composed of interlocking fetal and maternal tissues that function in the exchange of gases and nutrients between the circulatory systems of the mother and baby. An **umbilical cord** containing a substantial blood vessel system differentiates in the embryo by the end of the fifth week after implantation in the uterus. The embryonic heart begins pumping blood, and the whole mechanism for nutrition becomes operational. Wastes move out of the fetus and nutrients move into it across the placental membranes that separate the circulating blood of mother and fetus. There is no direct mingling of these blood systems in humans.

The placenta serves as the kidney, gastrointestinal tract, and lungs for the developing fetus, and also secretes hormones that are essential for the successful maintenance of pregnancy. The organ is formed predominantly from the **chorion**, or outer layer of embryonic membranes of the fluid-filled sac in which the fetus floats, but some portion of the placenta develops from the **allantois**. The allantois is an important and relatively large structure in reptiles and birds, but it may be quite small in many mammals. In humans and other primates the allantois is a small, hollow structure that projects into the umbilical cord at the site of its attachment to the fetal abdomen. The lining cells of the allantois contribute to the formation of the urinary bladder in the fetus, and allantoic connective tissue in primates gives rise to placental blood vessels. Thus, while most of the human placenta is derived from the outer chorionic membrane of the embryonic sac, its organization comes in part from the allantois as well (Fig. 6.2). Although the details vary among different species of mammals, the human placenta is torn from the uterine lining and is shed just after birth of the infant, along with some maternal tissue in the "afterbirth" material. This presents some slight disadvantage to the human mother since traumatic and physical damage can result from hemorrhage and torn tissue when the placenta is expelled.

Hormonal Changes During Pregnancy

The specialized structures and functions of the uterus during pregnancy depend upon the presence of high concentrations of circulating estrogen

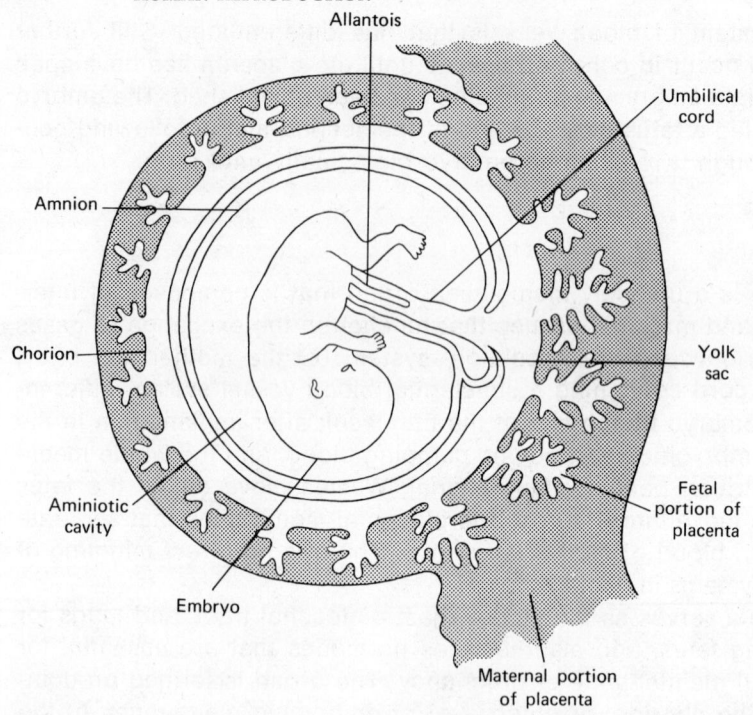

figure 6.2
Relationship of embryonic and maternal tissues in the development of the placenta.

and progesterone. Almost all of these steroid hormones are supplied from the active corpus luteum, which continues to develop in the ovary after ovulation if a successful pregnancy has been established. Otherwise the corpus luteum disintegrates within two weeks after ovum release, that is, by the time the next menstrual cycle begins. The continued presence and growth of the corpus luteum is essential for the continued secretion of estrogen and progesterone, these hormones acting to sustain the uterine lining. Menstruation, of course, does not occur during pregnancy.

Steroid secretion by the corpus luteum is strongly stimulated by **human chorionic gonadotrophin** (**HCG**), a hormone that is manufactured by the placenta. Although the two reproductive hormones are chemically distinct, HCG and luteinizing hormone (LH) have very similar properties. Since the blood levels of LH (and FSH) are very low during pregnancy, due to feedback inhibition when estrogen and progesterone levels are high, the presence and activity of HCG is important in encouraging high rates of secretion of the luteal hormones. Unlike their strong inhibiting effects on pituitary LH and FSH, the steroid sex hormones inhibit HCG secretion by the placenta to a much lower extent. Of course, there is no additional development of follicles, ovulation, or menstrual cycles during pregnancy because FSH and LH concentrations in the blood remain very low.

Many of the earlier tests used to detect pregnancy were based on the LH-like action of HCG. One such test, the Aschheim-Zondek or **A-Z test**, invented in 1928, was based on the principle that the presence of HCG in the urine (where it is excreted) of a woman in the second to fourth weeks of pregnancy was sufficiently high to induce ovulation when injected into test animals such as rabbits, frogs, or toads. In another test, male toads would shed sperm if the blood or urine sample contained sufficient HCG. Although tests using live animals yielded results that were 98 percent accurate, their relatively high cost and the time required (five days for the A-Z test) for the tests have mitigated their continued widespread use. More importantly, there now is available a very rapid and sensitive immunological test through which the presence of HCG in a sample can be determined while the patient waits for the results in the physician's office.

The rapid increase in HCG secretion reaches its peak about 60 to 80 days after the last menstrual period. After this, the rate decreases so that there is a very low but detectable level of HCG by the end of the third month of pregnancy, and its concentration remains fairly constant during the remaining six months. The reduction in HCG secretion is associated with marked increases in estrogen and progesterone secreted by the placenta, which continues for the final six months of pregnancy (Fig. 6.3). The persisting corpus luteum continues to produce some steroid hormones, but the preponderant secretion comes from the placenta. As a matter of fact, there is no adverse effect on the pregnancy even if the ovaries (containing corpus luteum) are removed after the end of the third month. If the ovaries should be removed during the first three months, then the fetus is lost (spontaneous abortion) almost immediately.

Development of the Embryo and Fetus

The period of time between fertilization of the egg and the birth of the newborn is called **gestation**. The duration varies among species (Table 6.1), but usually it is long enough in mammals to ensure the survival of the newborn without serious damage to the mother. Newly born rodents are hairless, unseeing, and unhearing so that considerable maternal care and protection are required at their birth. Mammals such as horses, cattle, deer, and others produce young that are able to run and engage in many activities within hours after their birth. Newborn dolphins can swim

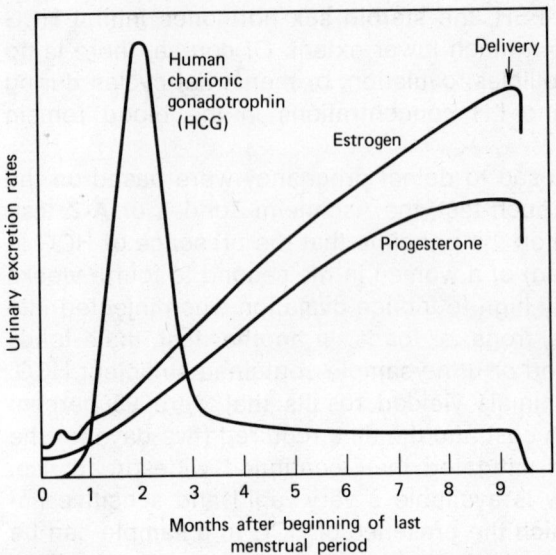

figure 6.3

Changes in hormonal levels during pregnancy. The secretion of human chorionic gonadotrophin (HCG) by the placenta is high during the first two to three months and then decreases to a low but constant level until delivery. The estrogen and progesterone secretion occurs principally in the ovarian corpus luteum during the first three months, but much higher amounts are produced for the final six months by the placenta.

table 6.1

Length of Gestation in Some Mammals

Nonprimate Species	Days	Primates	Days
Mouse	20	Lemur	120–140
Rat	22	Tarsier	180
Rabbit	31	Wooly monkey	139
Cat	63	Rhesus monkey	150–180
Dog	63	Chacma baboon	180–190
Lion	110	Langur	170–190
Goat	150	Gibbon	210
Domestic cattle	278	Orangutan	220–270
Horse	340	Chimpanzee	216–260
Dolphin	360	Gorilla	250–290
African elephant	660	Human	267, avg.

(From R. J. Harrison and W. Montagna, *Man*, 1st ed., p. 258. © 1969, Meredith Corp. Used with permission of Appleton-Century-Crofts.)

expertly within minutes. The human infant depends on parental care for the longest time of any animal, taking about one year to learn to walk and months just to establish sufficient motor control to be able to hold up its head. But, by the age of two or three years, the human child has stored away a phenomenal amount of learned information and a substantial repertory of abilities, including the uniquely human trait of symbolic communication using language.

During the 38 weeks of pregnancy, between conception and birth, the human embryo undergoes a profound series of developmental and growth changes that permit it to adapt almost immediately to the trauma of birth. At this time it is expelled from its warm, watery, and silent environment in the uterus; and it must secure oxygen, food, and other essentials by different routes than the umbilical connection to its mother.

The First Eight Weeks: The Embryo

The early phases of embryonic development comprise the most dramatic changes, beginning with the increase in cell numbers during the first week when the fertilized egg and subsequent embryo travel to the site of implantation in the uterus. The first cell division occurs about 36 hours after fertilization, to produce a two-celled stage. The two cells divide to make four at about 60 hours; the four cells divide to make eight by about 72 hours. These divisions result in an increased number of cells, but the whole structure is not much larger than the single egg cell from which it originated. Blastocyst formation takes place, with one side of the ball of cells being set aside as the future embryo and the other cells eventually differentiating into the membranes that will enclose the embryo. Until the sixth or seventh day after fertilization, the embryo derives its nourishment from the stored nutrients that were in the cytoplasm of the egg cell. At this time, the blastocyst makes contact with the uterine lining and initiates the chemical and cellular changes that permit the embryo to become embedded there. The embryo becomes surrounded by nutrients in the blood that escapes from the damaged endometrial cells, and it derives nourishment by absorption of this material. The cells of the blastocyst surrounding the embryonic cell-mass constitute the membrane primordia that develop into the outer **chorion** and the inner **amnion** membranes that contain the embryo within. The chorion develops numerous fingerlike projections that invade the maternal uterine tissues and, ultimately, a placenta comprised of embryo and maternal components will be formed (see Fig. 6.2).

Implantation continues during the second week, and the embryo gradually becomes embedded in the endometrial lining of the uterine wall. The embryo cell-cluster differentiates such that an **amniotic cavity** forms above it and some of these cells form a distinct **yolk sac** by the end of the second week. This sac is analogous to the structure in eggs of reptiles and birds, but no yolk is present in the yolk sac of the placental-mammal species. At this time the divisions and movements of cells lead to an embryo organized into a three-layered disc, wherein each of these layers of cell primordia carries a different potential for tissue and organ development. The embryo is about 1.5 millimeters long at the end of the second week, and its major body axis begins to develop (Fig. 6.4).

During the third week most of the major organ systems begin to form, the first of which is the beginnings of the central nervous system (brain and spinal cord). Two ridges form with a **neural groove** between them. Within the neural groove, the **notochord** and **nerve cord** develop, these two structures being characteristic of all animals of the great group of **chordates**, chief among which are the vertebrate animals. The notochord of the embryo is replaced by the bony vertebral column, and the nerve cord undergoes the first changes that will later lead to the differentiation of the brain and spinal cord. During this third week the rudiments of the heart, blood vessels, gut, and muscles also begin to form. Even though most women may not know or may not be sure if they are pregnant at this time, it is one of the most critical times of embryo development. Noxious chemicals, drugs such as thalidomide, irradiation, and mild rubella infection, among other things, may produce severe and permanent mental and physical damage to the embryo at this time, if it survives such exposure.

The eyes begin to form by 21 days; and approximately 100 cells are set

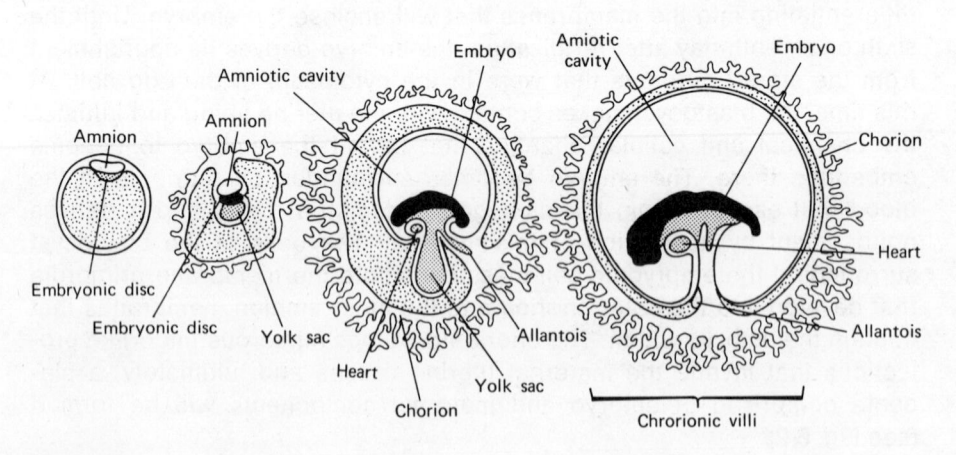

figure 6.4
Formation of the embryo during the first weeks of pregnancy.

aside in the yolk sac, as germ cells, from which the sperm or ova later will be produced. The embryo grows from a length of about 1.5 millimeters to about 5.0 millimeters in the fourth week, during which time the embryo changes shape from that of a flattened disc to form an enclosed "C"-shaped cylinder; the yolk sac shrinks until it forms the gut to which the allantois is attached at the hind end; and the embryo grows away from the surrounding chorion, but remains attached to it by a short body stalk that develops in all placental mammals into an umbilical cord, at a later stage. Specific and profound changes occur every day during the fourth week, leading to the laying down of the major organ systems or of their primordia. The tube that represents the extremely rudimentary heart begins to pulsate on the twenty-fourth day and continues to beat about 100,000 times per day until death stills this beat. The neural groove closes by the end of the fourth week; a definite tail is present; the head is proportionately as large as in other mammals; and forty paired segments of tissue are arranged lateral to the notochord. These 40 pairs of embryonic tissue segments will develop later into connective tissue, muscles, and bones. At the end of the first month of development, the embryo bears no resemblance at all to a human being, but it has reached a length of about 1/8 inch and has increased in mass about 7000 times (Fig. 6.5).

The heart begins to enlarge and supply blood to the embryo, as well as to undergo developmental changes that convert the tubular structure to

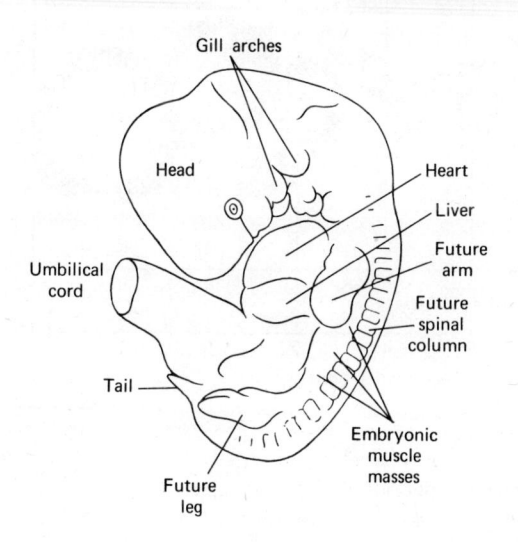

figure 6.5
Drawing of a human embryo during its fourth week of development.

134 HUMAN REPRODUCTION

a four-chambered heart by the end of the sixth week. The upper limb buds appear first and then the lower limb buds; and the tail reaches its greatest length just before it begins to regress during the second month. The brain begins to enlarge, eye and outer ear rudiments appear, and a pair of widely spaced pits mark the nostrils of an air-breathing animal. The head is noticeably large during the sixth week, and the nose and upper lip form from facial processes. The forelimbs develop to resemble arms with flat hand regions; furrows corresponding to the fingers will develop later. The hind limb buds undergo similar developments a few days later. The more compact mass of the embryo becomes altered in form, and buttocks appear as the tail becomes more inconspicuous.

The human embryo is about one inch long by the beginning of the seventh week and weighs about 1/28 of one ounce (one gram), but all of the major organ systems and body parts either are distinct or will be by the end of the second month (Fig. 6.6). The face forms during the seventh and eighth weeks, including the chin that is so characteristic of *Homo*

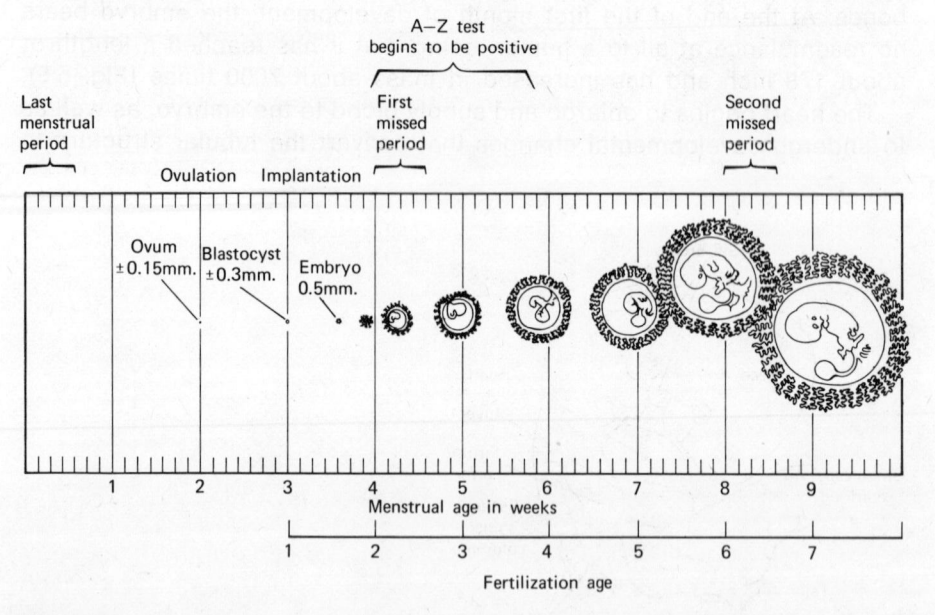

figure 6.6
The actual sizes of embryos and their membranes in relation to the time sequence according to the mother's menstrual periods and the age after fertilization has occurred. (From B. M. Patten, *Human Embryology*, 3rd ed., p. 145. © 1968 by McGraw-Hill, Inc. Used with permission of McGraw-Hill Book Company.)

sapiens among the primates, and fingers and toes become evident. The embryo has a bulging appearance due to the relatively large heart and liver. The liver alone accounts for about 10 percent of the body, and it functions as the main blood-forming organ of the fetus.

When this phase of development ends on the fifty-sixth day, the embryo has a relatively large head and brain, no external indications of its sex, and it measures about 1.5 inches in length. From this point until birth some seven months later, it is called a **fetus** and undergoes various changes, principally growth.

The Third Through the Sixth Months

As measured from the top of its head to its buttocks (sitting height), the fetus grows to a length of about three inches during the third month, and increases in weight to 1/2 ounce. Its body is arranged in the characteristic position of head bent forward, arms folded across the chest, and knees drawn up against the abdomen. The face becomes more apparent; nails form on the fingers and toes; and the prints of the fingers, palms, and toes develop well enough to be easily distinguished by ordinary methods of fingerprinting. The soft and cartilaginous skeleton begins to develop to bony structures, a process that is not completed until after puberty. The external genitalia begin to differentiate, providing the first external signs of the sex of the fetus; the kidneys and other excretory system structures develop quite rapidly at this time, as do the lungs, but all of these remain nonfunctional until after birth. The fetus can move its arms and kick its legs during the third month, and it shows such reflexes as sucking and swallowing. There is an apparent expressiveness to the face, such as frowning and squinting. By the end of this first trimester of development, all of the major organ systems have been laid down.

The fetus continues to grow and reaches a sitting height of about five inches and a weight of approximately five ounces by the end of the fourth month (Fig. 6.7). The lower abdomen and pelvis develop, the bony skeleton is forming, skin gland precursors develop, and the eyes and ears continue to differentiate. The first signs appear of the protective, cheesy coating that will increase in amount and cover the body of the fetus until it is delivered. This coating protects the fetus against abrasions and helps to maintain a constant body temperature. Fetal movements become quite obvious during this fourth month, and the uterus continues to expand as the fetus grows in size.

During the fifth month the fetus attains a weight of about half of one pound and measures about seven inches in sitting height. The beating heart is audible with a stethoscope. There is hair on top of its head and a

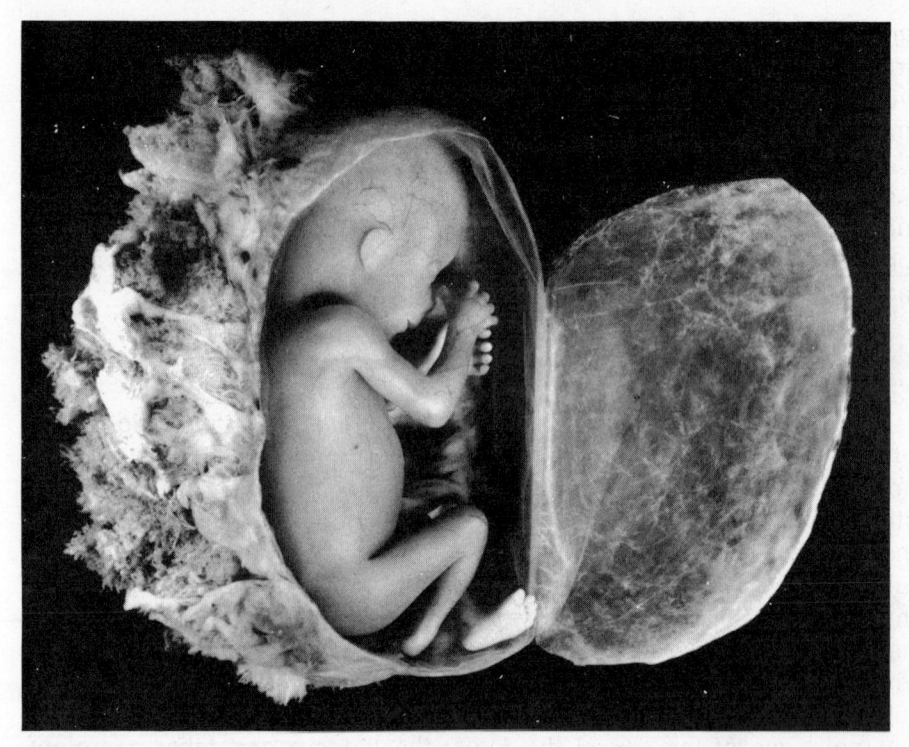

figure 6.7
A 16-week-old fetus shown within the fetal membranes (the chorion flap has been cut away to observe the fetus within). Photograph courtesy Carnegie Institution of Washington.

fuzzy down covers the body; the skin begins to lose its tautness and assume a wrinkled and folded look; further skeletal development takes place; and tooth development continues. Although fetal movement begins earlier and may be detectable in some women, most pregnant women can feel the spontaneous muscle movements of the fetus early in the fifth month; these movements becoming progressively stronger with time. A five-month-old fetus is unable to survive outside of the uterus. The youngest fetus known to have survived premature birth was about 23 weeks old, and it survived only because continuous assistance was provided for its vital functions of breathing, eating, and maintaining its body temperature in an incubator.

About 90 percent of the weight gain takes place during the sixth month of development of the fetus. It weighs about 1.5 pounds and has a sitting height of 12 to 14 inches. Its red and wrinkled skin is covered by an abundance of the cheesy coating, and there is a pasty mass of dead cells and bile in its intestines, which is not eliminated until birth. Fetal reflexes become more vigorous by the end of the second trimester of development.

The Final Trimester

The fetus doubles in size during the last two months of pregnancy, but its rate of growth begins to slow down during the ninth month. The fetus has a reasonably good chance of surviving premature birth at 28 weeks and afterward, especially with help in maintaining its body temperature. The fetus gradually becomes pudgier as the skin thickens and subcutaneous fat is deposited. Its eyelids are open at eight months, and it can perceive light and taste sweet substances. Although the ear apparatus is fully developed, the fetus cannot perceive sounds because there is no air available in the watery confines of the amniotic cavity, and thus no vibrations are possible in the bony structures of the middle ear. The fetus is well rounded at nine months and has a well-formed body and limbs. Hair on the head is prominent, as are the fingernails, and the testes in males have begun their downward descent into the scrotum, or even may be found there in some premature newborn boys.

During the last two to three months there is a rapid rate of formation of new brain cells and other parts of the central nervous system. There is abundant evidence showing that an adequate intake of proteins by the mother is essential for the full development of the nervous system of the fetus and hence of the intelligence of the child. In cases of severe protein deprivation in the mother, the child may be mentally less developed to some degree, a condition from which it does not fully recover regardless of its later diet.

The fetus usually acquires some antibodies from its mother's bloodstream during the ninth month, so that the baby is born with some level of immunity to whatever agents (bacteria, viruses, etc.) its mother is protected against. Additional immune substances can be transmitted through the milk to the nursing baby. These immunities are important because the newborn child does not begin to manufacture its own antibodies for about one or two months after birth.

The newborn infant at full term of 38 weeks (average of 266 days after ovulation, or conception; or, 280 days after the last menstruation) generally weighs about six or seven pounds and measures 19 to 21 inches from head to heels. Variations around these averages obviously occur. A series of striking changes takes place at birth, especially in the lungs, heart, and blood system. The respiratory tract and lungs begin to open as the baby takes its first breath, and this continues for several days after birth. Crying aids this process if it is sufficiently vigorous. When the umbilical cord is tied off and cut, a series of shunts takes place such that the pulmonary and body circulations become completely separated pathways. If these vascular closures do not occur, then the oxygen-poor blood of the veins mixes with the oxygen-rich blood of the arteries, and a "blue baby" results. The child may die if the condition is severe.

Difficulties During Pregnancy

Some physiological changes are quite evident in the pregnant woman, such as increased appetite and metabolic rate that result from the metabolic demands made by the growing fetus. A pathological condition known as *toxemia of pregnancy* may result from improper salt and water balance in the mother-to-be. If there is an excessive increase in salt and fluid in her body, along with hypertension and unusually high amounts of protein in the urine, then toxemia is indicated. The symptoms may be so severe as to cause convulsions in some women. One obvious contributing factor to this clinical condition is the retention of high levels of salt in the body. If these symptoms do develop, the treatment that is generally recommended is that of a salt-restricted diet. There is conflicting evidence concerning the efficacy of this diet, but treatment and therapy remain a matter of some guesswork since the causes of toxemia are largely unknown.

During pregnancy a great many substances normally pass across the placental "barrier" of fetal and maternal tissues, including gases, water, salts, hormones, nutrients for fetal growth and development, and all of the waste products of fetal metabolism that are returnable to the mother's circulatory system for disposal. Among the noxious substances that may diffuse or be transferred through the placenta to the fetus are various drugs, alcohol, and anesthetics. Some of these chemicals can lead to variable levels of damage to the fetus, depending upon the time during gestation (since the development of different embryonic tissues and organs occurs at different times) and the concentrations of the chemicals. Babies born of mothers who are addicted to narcotics show the agonizing symptoms of withdrawal and they suffer considerably, when they survive. Such babies often are born to women who neglect their own diet and rest, and who probably receive little or no medical care during pregnancy. These factors enhance the probability that the baby will not be particularly strong or healthy, and will be less able to overcome the drug-induced effects. Women who took thalidomide tranquilizers during early pregnancy gave birth to children with defective or missing limbs. The chemical exerted a deforming effect on the developing embryo, but its precise mode of action is not known, nor are the effects of a variety of such deforming substances.

The placenta usually provides an effective barrier to the passage of bacteria from the maternal tissues and circulation. Some viruses or their toxic products do cross the placental membranes and affect the developing embryo, especially during the first two months of pregnancy when many embryo organ systems begin to differentiate. If the mother contracts German measles (rubella) during pregnancy, the development of eyes, ears, and other embryonic rudiments may be defective. Blind or deaf children may have been born to mothers who had become infected with the rubella virus during early pregnancy, and a significant number of afflicted children might be born in a community where rubella assumed epidemic proportions at one time or another. Great care is exerted now if infection with the rubella virus is suspected to be possible, but until the unfortunate correlation was discovered there was no particular reason for pregnant women and their families to become overly concerned about a disease as innocuous as German measles. The effects on the embryo, however, are profoundly different.

The effects of some chemicals on the embryo or fetus may not become evident until after the baby is born and later in its life. There is some evidence showing that small doses of cortisone may cause cleft palate, and that the steroid diethylstilbesterol (DES) may be responsible for the development of vaginal cancer 18 to 20 years after birth in an individual whose mother was exposed to this chemical during pregnancy. This steroid has been added to beef cattle feed, and there is some question about its safety since we know little about the concentrations and persistence of the additive in beef products. At the moment, caution has been suggested. There are frequent reports from the laboratories of the Food and Drug Administration that lead to suggestions for caution and, in some cases, to the banning of sales of the questioned items and additives. Even though the evidence may not be conclusive, an item may be put on the restricted list if there is enough suggestive evidence of possible harmful effects to the consumer.

The developing fetus is exposed to immunity stresses when there are incompatibilities between maternal and fetal blood factors. One of the best known of these situations involves the Rh factor, most simply described as Rh-positive (factor present) and Rh-negative (factor absent) even though the inheritance pattern indicates greater complexity. The people who carry one or two Rh-positive alleles of the gene are capable of manufacturing Rh-antigens that circulate in the bloodstream along with other immune substances and factors. Individuals who are Rh-negative are recessive and thus carry only the Rh-negative allele on each chromosome of the pair where the gene is located, and they do not manufacture the Rh-antigen. If an Rh-negative mother carries an Rh-positive fetus (the allele derived from the father), then antigens produced in the fetus pass through the placenta and stimulate the production of Rh-antibodies by the mother (Fig. 6.8). These antibodies pass in turn into the fetal bloodstream, through the placenta, and may lead to severe effects because of antigen-antibody coagulating reactions. The effects usually are more severe in later pregnancies because the mother's immunity or sensitivity (antibody level) increases with each exposure to Rh-antigens during pregnancies when she carries an Rh-positive fetus. Eventually, there may be sufficiently high levels of antibodies to induce a crisis or death of the fetus or newborn child. Modern hospital procedures permit operations on the newborn so that a complete blood transfusion can be performed to replace entirely the defective blood, which is high in antibodies, with compatible blood that lacks these damaging components. There would be no reason for antibodies to develop later in such a child since its own body manufactures only antigens and not antibodies of the Rh variety. Similar immunological incompatibilities may result in response to one or another of the major A-B-O blood group factors, but the effects are not as dramatic or as serious as Rh incompatibility.

There is a much lower frequency of Rh-negative individuals among Negroes and Asians than in white Americans and Europeans. For this reason, the problems of Rh incompatibility are encountered more often by whites since their populations include many more Rh-negative individuals against a background of predominantly Rh-positive people. The chances for pairing between Rh-negative women and Rh-positive men thus are higher among whites than in other racial groups. There are no

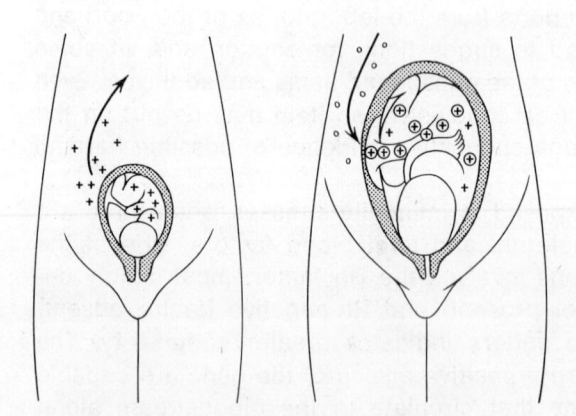

Rh—negative mother and Rh—positive fetus

figure 6.8
Relationship between Rh-positive fetus and Rh-negative mother. Antigens from the fetus enter the mother's circulation, where antibody formation is stimulated. The circulating antibodies enter the fetus through the placenta and agglutinate the resident antigens that may lead to fetal blood problems.

complications due to the Rh factor when the mother is Rh-positive, regardless of the Rh type of the fetus, because she does not produce Rh-antibodies, and the fetus is incapable of manufacturing antibodies of any kind. Indeed, the capacity to manufacture immune substances does not develop for one or two months after birth.

Birth

Labor

Birth of the baby, known as **parturition**, is accompanied by an interval of **labor** by the mother. Labor occurs in three stages: *dilatation* or opening of the cervix in response to strong and rhythmical uterine contractions; the *expulsion* of the fetus begins with the appearance of its head in the dilated cervical opening; and finally, the *placental* stage that involves voiding of the placenta with the attached severed umbilical cord after the delivery of the baby (Fig. 6.9). The first stage lasts from about 2 to 24 hours and usually is longer for the first baby than for subsequent births. Weak and infrequent uterine contractions actually begin as early as 30 weeks of pregnancy, and these become stronger and more frequent during the last two months. The entire uterine contents shift downward during the ninth month, and the fetus is brought into contact with the cervix. The baby's head acts as a wedge that further dilates the cervical opening to the vagina, or birth canal. The mild uterine contractions occur every 15 to 20 minutes at first, but increase in strength and come about every one or two minutes toward the end of this first stage of labor. Prosta-

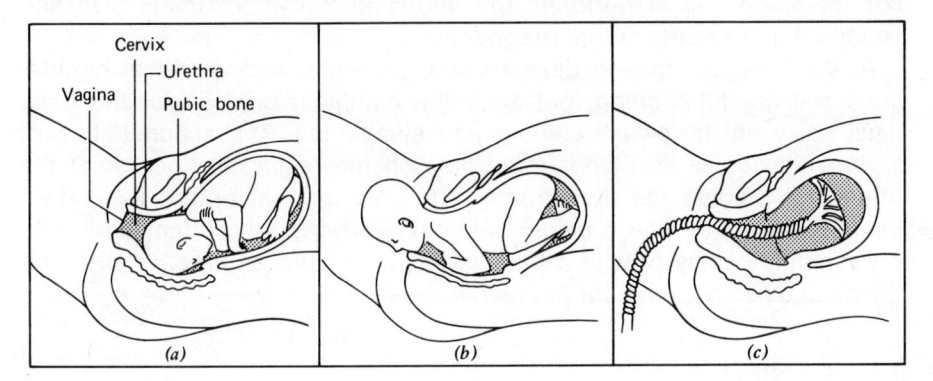

figure 6.9
The three stages of labor at birth: (a) dilatation of the cervix, (b) expulsion of the fetus, and (c) shedding of the placenta and remainder of the umbilical cord, plus some uterine tissue.

glandins are found in the amniotic fluid and in the venous blood of women undergoing the contractions of labor, and these fatty-acid substances probably induce or increase uterine muscle contractions. If prostaglandins are supplied during a difficult labor, delivery may be induced within a few hours. This aid was not available a few years ago.

During the dilatation stage, the membranes surrounding the fetus are ruptured and intrauterine fluids are expelled through the cervix and vagina. Uterine contractions force a wider cervical opening (about 10 centimeters, or 4 inches), and the fetus moves from the uterus through the cervix into the vaginal canal under the influence of hormonal stimulation of contraction. If the mother acts to increase the ongoing abdominal reflex by "bearing down," then this provides a further aid to the movement of the baby by uterine contractions. The delivery is easier if the baby's head is oriented downward, which occurs in 75 to 90 percent of the cases, since the head is its widest part (about 5-1/2 inches). The greater dilatation of the cervix and vagina permits the subsequent passage of the remainder of the baby's body to occur more readily. If there is a breech birth, then there may be some distress to the infant since the umbilical cord may become constricted if it gets caught between the baby and the birth canal after the lower portion of the body has emerged first.

The expulsion stage of labor may last only a few minutes or even an hour. The uterine contractions occur about one or two minutes apart and last for a minute or so. After the baby is born and the placental stage is initiated, the placental blood vessels contract and the placenta is released from the uterine wall. Uterine contractions continue for a while until blood, fluid, placenta, and some maternal uterine tissue are expelled as the "afterbirth." Minor contractions continue, and these aid in stopping the flow of blood and in returning the uterus to the approximate size and condition that existed before pregnancy.

At the time the baby is delivered the placental and umbilical circulations still are functioning, but once the umbilical cord is cut the baby must carry out its own breathing and elimination. The accumulation of carbon dioxide in its bloodstream acts on the respiratory center of the brain and triggers the first breath. The cold air that enters the baby's lungs for the first time provides a shock, and once the infant begins to cry (spontaneously or with a helping slap) it begins to breathe and continues to do so regularly for the rest of its life.

Multiple Births

Multiple births generally present fewer difficulties in other mammals either because the fetuses are arranged in a line within the uterus, or because the body of the fetus is arranged more compactly, or because of a difference in the nature and arrangement of the fetal membranes. About one pregnancy in 85 results in twins, with a slightly higher rate of twin births among Negroes than whites contributing to this overall average for the United States population. Two-egg twins occur twice as often as twins produced from a single egg that later divides to produce two identical embryos. Twins that develop from a single fertilized egg are always of the same sex and are genetically identical since the same combination of parental chromosomes is present in both. Twins that result from two eggs, each fertilized by a different sperm, may be of the same sex or of opposite sexes depending on the kinds of sperm that fertilized each egg. As you might expect on the basis of chance, about 50 percent of two-egg twins are the same sex and about 50 percent are brother and sister.

Twins produced from one fertilized egg seem to be born to mothers of all ages, whereas older mothers are more likely to produce two-egg twins. There probably are genetic factors that influence two-egg twinning since the trait often runs in families; women from high-twinning families are more likely also to have twins. Environmental factors almost certainly influence twinning, too, and it has been observed that the frequency of twinning has declined recently in some countries, including the United States.

Hormonal Physiology

Some of the stimuli and chemical interactions in parturition are known but, like so much of human reproductive physiology, there are large gaps in our knowledge. We actually know more of the details of reproductive physiology in some kinds of domesticated animals than we know about ourselves. This distortion is due to many factors, some financial and others concerned with the accepted lines of research permitted to medical and nonmedical professionals in studies of human biology.

Shortly before delivery the uterine smooth muscle is activated by prostaglandins and an increase in estrogen, but the progesterone level decreases at the same time. The decrease in progesterone relieves inhibition of myometrial contractions. Uterine contractions proceed under the influence of a neurohormone, **oxytocin**, as well as prostaglandins and estrogen. Oxytocin is a very small molecule made up of only eight amino acids, and it is secreted by special nerve cells in the hypothalamus portion of the brain but is stored in the posterior lobe of the pituitary gland. In response to signals from the uterine and cervical regions, oxytocin is released from the pituitary storage cells and is transported through the bloodstream to the uterus, where it aids in stimulating contractions during labor.

The onset of labor involves a coordination of nervous, hormonal, and muscular activities that continue to operate to distend the uterus and aid in the expulsion of the fetus. Animals still will undergo "labor" even if the fetus has been removed some weeks earlier, which provides evidence that neither uterine distension nor the presence of a fetus is essential to the events of labor.

Lactation

Anatomy and Physiology

The development of a girl's breasts is induced primarily by ovarian estrogens. The breasts begin to develop before puberty, but begin to enlarge visibly at puberty and continue to grow until late adolescence. Most of the breast is comprised of fat and connective tissues, with relatively little mammary tissue as such, regardless of the size of the breast. The milk ducts, which converge at the nipples, branch throughout the breast tissues and terminate in saclike glandular clusters that resemble grapes attached by stems to the ducts. Milk is secreted in these glandular clusters. Estrogen and progesterone influence the development of the mammary ductile and glandular tissues at puberty, along with prolactin and somatotrophic (growth) hormone, two proteinaceous hormones secreted by the anterior pituitary under the control of the hypothalamus in the base of the brain.

Breast enlargement begins fairly early in pregnancy and continues until term. The combined action of placental estrogens and some anterior pituitary growth hormone leads to the stimulation of functional mammary tissues and an accumulation of surrounding fat and connective tissue. Maximal development also requires the simultaneous action of progesterone. Small amounts of a clear fluid are produced near the end of pregnancy which resemble milk except for the virtual absence of fats. True milk production begins shortly after the baby is born, stimulated by **prolactin** that is secreted in the anterior pituitary in higher amounts at this time. The secretion of this particular pituitary hormone is inhibited during pregnancy by the high levels of placental estrogen and progesterone, which diminish during labor and thus release the suppression of prolactin manufacture. Prolactin also is secreted in men, but its function, if any, is entirely unknown.

Lactation does not proceed spontaneously (Fig. 6.10). Removal of the fetus in experimental animals has no effect on inducing lactation, but removal of the placenta at any time late in pregnancy will lead to lactation. This can be explained primarily as the effects of progesterone secretion by the placenta, which enhances the synthesis of a prolactin-inhibit-

Fertilization, Pregnancy, Birth, and Lactation

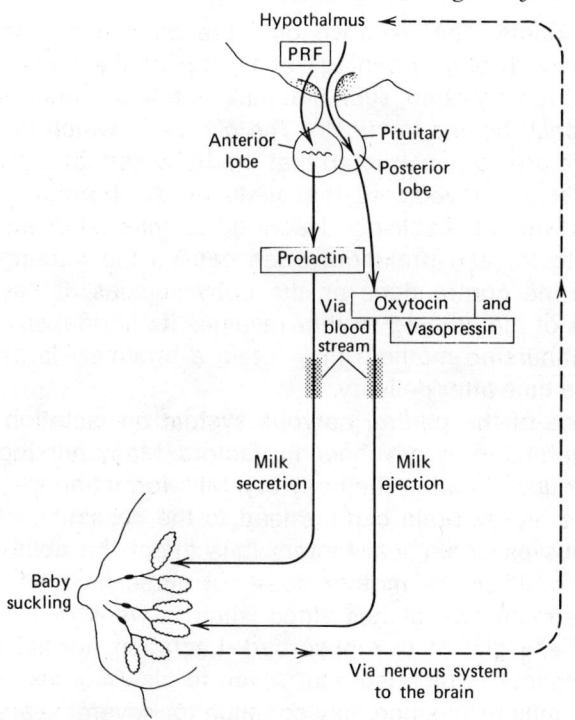

figure 6.10
Interacting systems involved in lactation, stimulated by the suckling of the baby. (PRF=prolactin releasing factor)

ing factor from the hypothalamus. An alternative explanation would be that progesterone prevents the synthesis of a prolactin-releasing factor in the hypothalamus, thus restricting pituitary release of prolactin itself. In either explanation, and we cannot decide between them, delivery of the baby and expulsion of the placenta removes the sources of progesterone and its inhibition of prolactin secretion. In addition to prolactin, another important factor that induces lactation is the stimulation of the nipples of the breast by the suckling of the infant. There is a continuous reflex input to the hypothalamus from nervous receptors in the nipples when the baby suckles, so that milk production may continue for years if nursing is maintained, but will stop very soon after the mother stops nursing the infant. The nerve impulses to the hypothalamus are then transmitted to the posterior lobe of the pituitary gland, which leads to the release of oxytocin along with a very similar neurohormone called **vasopressin**. The small musclelike cells around the enlarged mammary glands contract under the influence of these two neurohormones, so that milk is ejected from the glands into the ducts. The negative pressure caused by the suckling brings out the milk through openings at the tip of the nipple. The actions and reactions are very rapid, such that milk "let-down" occurs within a minute after the baby begins to suckle. The glands in which the milk is secreted and stored are so constructed that the baby cannot suck it out directly; it must first be moved into the ducts by the hormonal, nervous, and muscular activities collectively described as **milk let-down**. The reflex release of oxytocin (and prolactin) often causes the nursing mother to experience uterine contractions as the baby suckles. It has been noted that the uterus of the nursing mother resumes its normal size more rapidly, whereas nonnursing mothers may retain a relatively large and flaccid uterus for some time after delivery.

Because of the influence of the central nervous system on lactation, milk may be released or inhibited by psychogenic factors. Many nursing mothers begin to leak milk upon hearing the baby cry. Milk formation may stop if disturbances to the higher brain centers lead to the cessation of oxytocin release. Numerous psychological factors thus affect the ability of a woman to nurse her child. If the mother does not nurse her child, then milk secretion in the mammary glands stops within a week or two and any stored milk in these glands is reabsorbed. Lactation normally ceases after the child is weaned, which is about seven to nine months in most human societies; but milk production may continue for several years if nursing continues to be practiced.

Ovulation is inhibited in most women during the early months of lactation, and menstrual cycles do not occur during this time. Another neurohormonal reflex that is triggered by suckling apparently leads to the hypothalamic inhibition of the release of FSH and LH from the anterior pituitary, thus blocking ovulation (see Fig. 5.10). This inhibition does not persist for very long in many women so that about 50 percent begin to ovulate even though nursing is continued. Unplanned pregnancies occur often in women who believe that ovulation is prevented while active lactation continues. Although much more common in other primate species, some women may become pregnant a few weeks or months after giving birth and still continue to nurse the baby until the next infant is born.

The manufacture of milk is a rather complex and intricate series of processes. The raw materials (amino acids, glucose, fatty acids, etc.) must be extracted from the blood by the mammary glandular cells and synthesized into the major milk solids: protein, fat, and the carbohydrate *lactose* ("milk sugar"). A number of hormones and enzymes all must act coordinately in time and place before the milk product is manufactured. In addition to the water and solids that comprise the milk of the nursing mother, immune substances also are transmitted to the infant through the mother's milk. Since newborns do not develop their own immunity mechanisms for a month or two after birth, they derive some of these important components for disease resistance directly from their mothers. This observation provides one of the arguments in favor of nursing, but there are other personal reasons that may influence the mother to use bottled formula milk instead.

Lactose and Lactase

The lactose sugar molecule is made up of two smaller and simpler sugar units, one glucose and one galactose per lactose molecule. Lactose can be used in body metabolism to secure energy and materials for growth, but first it must be digested by the enzyme *lactase* to glucose and galactose units. Glucose and galactose then can be used directly by body cells. The human fetus begins to produce the enzyme lactase during the latter part of the gestation period, and the enzyme activity is at a maximum shortly after birth. The activity of the enzyme decreases in most mammals afterward, reaching low levels in most human children sometime between the ages of one and a half to three years.

In recent years there have been sufficient studies to indicate that most of the world population cannot digest the lactose in milk products, although the ability is retained into adulthood by some groups of people. Whereas almost every young child can digest milk sugar, lactase activity is low or absent in East Asians, most Africans and South Americans, American Indians, Eskimos, and a substantial proportion of southern Europeans. The principal population groups who retain lactase capacity are the groups from northern Europe (and many Americans) and certain pastoral African tribes such as the Tussi of Uganda and the Fulani of Nigeria. About 90 percent or more of other population groups are deficient in lactase, and from 6 to 15 percent of Americans and northern Europeans are lactase deficient. American blacks probably showed higher frequencies of lactase deficiency in earlier centuries, but the incorporation of a substantial amount of Caucasian genes undoubtedly explains the present proportion of approximately 70 percent who cannot metabolize lactose adequately.

From some genetically oriented observations of different human groups, it appears that tolerance of lactose is dominant whereas intolerance behaves as a recessive trait. Since most of the world cannot metabolize milk sugar after early childhood years, this provides one example of recessive alleles for a particular trait occurring in higher frequencies than the dominant alternatives.

The difference in ability to digest lactose is reflected in the absence of milk-based sauces in some cuisines, such as Chinese, Indonesian, and other Asian groups. The more frequent use of milk in creamy sauces of western cooking obviously indicates a high level of lactose tolerance in these peoples. The hearty American breakfast that includes cereal in milk, milk in coffee, and a tall glass of milk, may serve us rather well nutritionally. But this reverence for milk as a universal nutritive food unfortunately has shaped our policies in providing food to people of other countries. Powdered milk is shipped easily, but it causes gastrointestinal distress and great discomfort to children and adults of most of the world. It is shunned strenuously by the people who have had the unfortunate experience of ingesting gift milk. Cheeses and fermented milk products, such as yogurt, can be digested readily, however, since most of the lactose has been converted to glucose during the processing of the raw milk.

Suggested Readings

Allen, R. D., The moment of fertilization. *Scientific American, 201* (July 1959), 124.

Barry, J. M., The synthesis of milk. *Scientific American, 197* (October 1957), 121.

Clarke, C. A., The prevention of "Rhesus" babies. *Scientific American, 219* (November 1968), 46.

Cooper, L. Z., German measles. *Scientific American, 215* (July 1966), 30.

Csapo, A., Progesterone. *Scientific American, 198* (April 1958), 40.

Etkin, W., How a tadpole becomes a frog. *Scientific American, 214* (May 1966), 76.

Harrison, R. J., and W. Montagna, *Man.* New York: Appleton-Century-Crofts, 1969.

Ingalls, T. H., The strange case of the blind babies. *Scientific American, 193* (December 1955), 40.

Ingalls, T. H., Congenital deformations. *Scientific American, 197* (October 1957), 109.

Kretchmer, N., Lactose and lactase. *Scientific American, 227* (October 1972), 70.

Lehrman, D. S., The reproductive behavior of ring doves. *Scientific American, 211* (November 1964), 48.

Patton, S., Milk. *Scientific American, 221* (July 1969), 58.

Pike, J. E., Prostaglandins. *Scientific American, 225* (November 1971), 84.

Pincus, G., Fertilization in mammals. *Scientific American, 184* (March 1951), 44.

Smith, C. A., The first breath. *Scientific American, 209* (October 1963), 27.

Taussig, H. B., The thalidomide syndrome. *Scientific American, 207* (August 1962), 29.

chapter seven

Fertility and Infertility in Humans

The global modern problems of overpopulation have stimulated vigorous searches for effective birth control methods and materials. Among the most ancient biological methods of birth control are methods that involve partial or total abstinence from sexual intercourse, as well as others that focus on primitive but effective mechanical devices. The infamous "chastity belt" that shackled the wife at home while her husband was away on a crusade or some other worthy cause provides a typical European Middle Ages touch to the history of sexual control... a combination of cruelty, parochialism, and nongallantry. The principal approach to fertility control before 1950, at least in the United States and western Europe, involved techniques and materials that prevented the sperm from reaching the egg. Since the development of oral contraceptives such as the "pill," there has been a swing away from these earlier methods, at least in the developed countries of the world. The means for achieving fertility or birth control can be subdivided into four main groups: sexual abstinence, contraception, sterilization, and abortion. We will discuss each of these briefly.

Methods of Fertility Control

Sexual Abstinence

The **"rhythm method"** depends upon the fact that ovulation occurs once each 28 days during the menstrual cycle, and that the shed ovum retains its viability for about 10–15 hours (perhaps 24 hours) after its release from the ovary. Since human sperm may remain viable for an average of 48 hours but as long as 72 hours after intercourse, abstinence generally must be practiced for at least three days during the menstrual cycle of the woman. If intercourse does not take place, then the unfertilized egg will degenerate with 24 hours or less after its release from the ovary. The greatest difficulty in relying on this method is that ovulation does not occur at the same fixed time in each menstrual cycle. Even with tests to measure slight increases in body temperature or of progesterone level as indicators of ovulation, there is no assurance that subsequent ovulation times will be fixed or invariant. Because of variations in the time of ovulation even within the same individual, there is a 25 percent or higher probability for conception by women who rely on the rhythm method for birth control. The odds of one to four that a pregnancy might result from intercourse practiced during "safe" times has encouraged many people to seek alternative methods for family planning.

The major timing variation occurs during the first part of the menstrual cycle leading to ovulation, whereas a more constant average interval of 14 days characterizes the *postovulatory* phase. Even for a regular 28-day cycle the egg may be released any time from the seventeenth to thirteenth day before the beginning of the next menstrual period. If there has been intercourse two days before ovulation, then live sperm still may be present in the reproductive tract and be capable of fertilizing the egg. If intercourse should occur even one day after the latest possible time for ovulation, then the egg might still be viable and a pregnancy could be established. In view of these figures for women with regular 28-day cycles, abstinence from coitus thus would be required from the tenth to the seventeenth day after the beginning of the last menstrual period. In such cases showing regular cycles, as many as eight days each month could be considered "unsafe" for couples who depend on the rhythm method. Since most women experience considerable irregularity in their menstrual cycles, and since it is not very practical to expect the next cycle to begin on a particular day in each calculation, the rhythm method leads to relatively frequent instances of unplanned conception. In addition, some recent evidence has indicated that ovulation may be induced on occasion by the stimulus of sexual intercourse itself. If this is true, then it explains some of the basis for variable duration of the preovulatory phase among women, and also of the high level of unreliability of fertility control when the rhythm method is utilized. Considering the disadvantages of the rhythm method and the low appeal for many people of managing their sex lives according to the calendar rather than personal feelings and desires, this alternative has been abandoned by those couples who can choose from among a number of other methods.

A life of **celibacy** involves total abstinence from sexual intercourse and is the designated or chosen mode for various religious groups. Priests usually are celibate in religions such as Roman Catholicism and Buddhism, but not in many other religions of either eastern or western origins. Women in religious orders usually refrain from sexual activities, even in those Protestant denominations that do not require celibacy of male clerics. It is a common double standard of behavior.

Sexual abstinence is a frequent mode of fertility control in times and places of overpopulation. From 20 to 60 percent of some groups of people in eighteenth- and nineteenth-century western Europe were celibate, either because of enforced situations imposed by employers or the government, or for economic reasons.

The technique of **coitus interruptus**, or **withdrawal**, is mentioned in the Old Testament scriptures. It is the usual method for birth control of Roman Catholics in European countries today, and it was a more common practice in the United States before the development of chemical and mechanical contraceptives. Coitus interruptus requires the male to withdraw his penis from the vaginal tract just before ejaculation is to occur. Successful withdrawal depends on the male since he must be aware of the first signs of approaching orgasm and also be prepared to terminate intercourse at that time. Since orgasm represents the time of greatest sexual satisfaction and intense pleasure, there may be psychological problems that could disturb the sexual relationships that depend on the technique of coitus interruptus during the "unsafe" times in the woman's menstrual cycle. In addition to these potential difficulties, surveys of the effectiveness of various birth control methods have indicated a considerable level of failure of this particular method. For every hundred couples using withdrawal for a period of one year, about 18 women are likely to become pregnant (Table 7.1). An important factor that leads to lowered levels of success for the withdrawal method is that there usually are some sperm present in the very first fluids released from the penis, even before orgasm occurs. Any mistake in the timing of withdrawal also creates a problem since one droplet of semen may contain enough sperm to effect pregnancy. In fact, there is a higher concentration of sperm in the first drops of semen that are discharged than in the seminal fluid released later during ejaculation.

table 7.1
Relative Effectiveness of Contraceptive Methods[a]

Method Used	Pregnancies per 100 Women per Year
None	115
Douche	31
Rhythm	24
Jelly alone	20
Withdrawal	18
Condom	14
Diaphragm	12
Intrauterine devices	5
Oral contraceptive: sequential steroids	5
Oral contraceptive: combined steroids	0.1

[a] Data from studies of G. Pincus and others.

Contraception

Birth control by **contraception** generally is accomplished in either of two ways: prevention of ovulation, or prevention of the union of egg and sperm. The methods that are used rely on mechanical, chemical, or hormonal factors.

Mechanical Obstructions

Of the mechanical devices, all of which serve to prevent fertilization, one is used by men and the other two kinds by women. The **condom** consists of a synthetic rubber sheath that is worn tightly over the erect penis. It can be even more effective if it is used in conjunction with a coating of contraceptive jelly on the outside of the condom, or if foam is introduced into the vagina before sexual intercourse begins. The condom may serve as a therapeutic means for reducing the spread of venereal disease, which may have been its original function when it was invented in the sixteenth century, as well as serving the primary purpose of preventing the entry of sperm into the vagina. The sheaths are readily available and are the most commonly used contraceptive device, even with increasing utilization of the contraceptive pill and of the newer intrauterine devices. The major disadvantages of the condom are purely personal in that it dulls the sensations that otherwise contribute to full enjoyment of the sexual act, and the man must put it on during the sexual relationship. This sort of interruption can be quite aggravating or esthetically unpleasant.

Some precautions are required if the condom is to be maximally effective. Prompt removal of the penis from the vagina is desirable to ensure that semen will not leak out over the top of the condom, and re-use of the same condom is impractical since washing may produce holes in the condom if it is used in repeated ejaculations.

The **vaginal diaphragm** or cap is designed to cover the cervical opening into the uterus so that sperm are blocked from entering the upper portion of the female reproductive tract. The diaphragm consists of a rubber or synthetic rubber cap with a stiffened rim of flexible metal coil or spring that holds the cap in place. The diaphragm coil comes in different diameters (two to four inches) to accommodate to various vaginal sizes, but it must be fitted to the woman by a trained person since it could be displaced or lost if it is the wrong size. Once fitted, the diaphragm usually is purchased by prescription at the drug store. The vaginal opening may become wider after childbirth or with time once sexual intercourse has become a regular part of the woman's activities, so that the prescription must be renewed at intervals. In addition to obstructing sperm entry, the diaphragm provides much greater insurance against conception when a contraceptive cream or jelly is applied around the edge and inside the cap. The sperm are killed on contact with the spermicidal chemical in the preparation.

The diaphragm must be inserted before coitus and must remain in place for 6 to 12 hours afterward, although it causes no harm or discomfort if it remains in place for as long as 24 hours after intercourse. The same cap can be used safely for several years if it is properly washed and dried after removal. This method is as disagreeable to the woman as the condom is to the man, and it is no safer to use. The relatively high rate of pregnancy (12 percent of the women who use it for one year) has contributed to the decline in the use of this particular mechanical contraceptive device.

Various **intrauterine contraceptive devices** (IUD, or IUCD) have an ancient history but have recently come into common use in countries such as the United States and Great Britain. These IUDs measure about one inch in diameter and consist of a ring of stainless steel or more usually are a spiral, loop, or bow made of plastic (Fig. 7.1). Because the IUD is aseptically inserted directly into the uterus rather than the vagina, and may remain there indefinitely, it must be put into place by a trained person. If pregnancy is desired at some later time, then the IUD is removed and the uterus becomes normally functional within six months to a year, after which pregnancy can take place. The precise mechanism by which the IUD functions really is not known, and its action is considered to be only partly that of a mechanical obstruction, if at all.

Because of individual problems and some very unpleasant side effects,

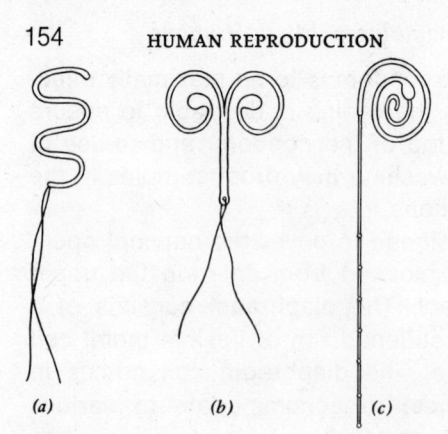

figure 7.1
Three commonly available intrauterine contraceptive devices: (a) Lippes loop, (b) "Saf-T-Coil," and (c) Margulies spiral with stem. The thread or stem projections indicate that the IUD still is in place and aid in its removal.

such as excessive bleeding and nausea, only about two-thirds of all women can use the IUD satisfactorily. This device is a more reliable fertility control since it reduces the rate of unexpected pregnancies to about five per hundred women using an IUD for one year. There is no need to remove the IUD if pregnancy should occur since it provides no interference with the normal development of the fetus or the delivery of the baby. Because it is relatively cheap and effective, as well as simple to have installed, the intrauterine device provides a useful alternative to other methods for population control. Many women in the developed countries have come to use the IUD because it is cheaper than the pill and it is not necessary to keep track of the number of days or to remember to take one each day. It should become more widely used in the developing countries of the world for these same reasons, and because oral contraceptives are not always available or in adequate supply. One major difficulty, however, is that the IUD may be passed out of the uterus without the woman necessarily realizing it. Where even one more pregnancy can cause a hardship on the family, this disadvantage must be considered carefully. If the IUD is to become more widely used in the world, then we must know more about the mode of action than is available at present.

Chemical Contraceptives

Chemical contraceptives include **douches**, vaginal **spermicides**, and **pills** that contain hormonal ingredients. The practice of douching after sexual intercourse is relatively futile since vaginal washes will not remove sperm that already have entered the uterus, perhaps within one minute after the ejaculate has been deposited in the vagina. The variety of spermicidal preparations also prove to be unreliable when used as the sole contraceptive method (about 20 percent annual rate of pregnancy). The unplanned pregnancy rate is reduced when the spermicide is used in conjunction with a condom or diaphragm, but many of the foams, creams, and jellies are designed to be used alone.

The **oral contraceptive pills** prevent ovulation and have proven to be highly effective in some forms. The principal component of the pill is the synthetic hormone *progestin* that has properties similar to those of the natural steroid progesterone. Although either of these chemicals would prevent pregnancy if administered regularly, progesterone loses its activity if it is swallowed whereas the synthetic hormone has been manufactured in a form that retains activity when taken orally. Most of the 30 or more brands of the pill also contain a modified form of the estrogen type of steroid hormone. The hormones in the pill act like the hormones in the pregnant woman, by suppressing ovum release. Ovulation is inhibited by high concentrations of progesterone and estrogen, which not only keep the uterus in a suitable state for pregnancy but which also act on the pituitary gland (see Fig. 5.10). The secretion of follicle-stimulating hormone (FSH) and luteinizing hormone (LH) by the anterior pituitary is inhibited by the ovarian progesterone and estrogen (see Fig. 5.11). FSH and LH in turn stimulate follicle maturation and ovum release. When hormonal production is blocked the ovary fails to release ova because follicles do not ripen under these conditions. This self-regulatory feedback control was discussed earlier (see Chapters 5 and 6). When the pill is taken daily beginning on the fifth day after the start of menstruation and continuing until the twenty-fifth day of the cycle, that is, for 21 days in succession, its contraceptive effectiveness reduces the rate of accidental pregnancy to about one per thousand women using the pill for one year. The woman stops taking the pill for the remaining three days of the menstrual cycle, during which time there is some bleeding but less than during ordinary menstruation. The 28-day cycle established with the pill is artificial, but it fulfills a physiological (albeit unnecessary) as well as a contraceptive function.

The first pills were introduced in the mid-1950s by Gregory Pincus and his colleagues, and they were of the *combined* type containing modified progesterone and estrogen components (Fig. 7.2). Two other forms of the pill have since been placed on the market, the *sequential pill* and the *minipill*. Depending on the particular brand, the sequential pills consist of a first kind that contains only an estrogen that is taken for about 15 days (beginning with day-5 of the established menstrual cycle), followed by the second kind of pill that has a combination of progestin and

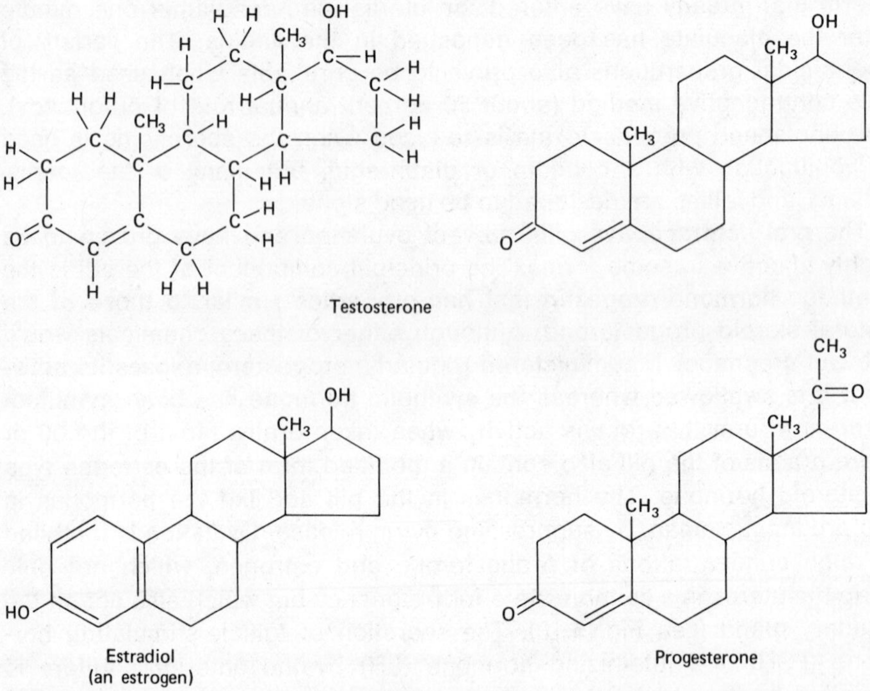

156   HUMAN REPRODUCTION

figure 7.2
Three of the naturally produced steroid sex hormones. The principal male androgen, testosterone, is very similar to the two major female sex hormones, estradiol and progesterone. The chemical formulae usually are shown in a shorthand notation, but the positions of the carbon and hydrogen atoms are indicated in the formula (of testosterone) in the upper left corner.

estrogen and that is taken for at least 6 days. The total duration of the contraceptive interval usually is still 21 days, but the absolute number of days for each kind of pill varies among the different brands. The rate of unplanned pregnancy rises to 5 percent with this method as compared with 0.1 percent when the combined pill is used exclusively.

The minipill contains no estrogen, which is the agent responsible for most of the undesirable side effects of the other pills, and it has only about one-fourth the amount of progestin of the other alternatives. It is taken orally every day and appears to be quite reliable since the annual rate of unplanned pregnancy is only about one percent. Rather than suppress ovulation, the minipill is believed to exert its hormonal effects on the uterine lining so that implantation is prevented. But this has not actually been demonstrated.

The oral contraceptives would appear to be extremely reliable, rather simple to use, and have no permanent effect on fertility since a pregnancy can be initiated when a woman stops taking the pill. But these materials alter the entire reproductive physiology of the woman. So little is known of the human system that some care must be taken in this regard. There is direct clinical evidence showing that some women develop a blood clot in the lungs after taking hormonal contraceptives, although it is quite rare and usually not fatal to the patient. What the physicians have termed "undesirable" side effects may be sufficiently severe or distressing to cause many women to avoid using oral contraceptives. Some evidence exists that oral contraceptives may have a cancer-inducing effect, but we cannot be sure about this possibility until at least 20 years worth of data have been collected and evaluated; which will be some time in the late 1970s or the 1980s. The estrogenic component of the pill causes nausea because gastrointestinal muscle contraction is induced; a "bloated" feeling because the hormone causes water retention; facial blotches may occur because of high levels of copper in the blood due to estrogenic oxidation of skin pigments; migraine headaches; backache; and other symptoms. The omission of estrogen does not help too much since progestin alone often causes excessive bleeding.

The interference with the whole reproductive system as well as with other parts of the body has stimulated considerable search for more effective and safer contraceptives than the pill. Various lines of study are now in progress, and some new approaches have been made possible by very recent discoveries. One approach takes advantage of the recent success in synthesizing the LH-releasing factor (LHRF) of the hypothalamus (Fig. 7.3). It is possible to chemically modify the synthetic LHRF and to produce molecules that are antagonistic to the natural LHRF. If such specific antagonists were used as contraceptives, they would perhaps prevent LH release from the pituitary and thus prevent ovulation since the ovum will not mature in the absence of LH (and FSH). There is

(pyro)Glutamic acid-Histidine-Tryptophan-Serine-Tyrosine-
Glycine-Leucine-Arginine-Proline-Glycine-NH$_2$

LUTEINIZING HORMONE RELEASING FACTOR

figure 7.3
Recent successful chemical analysis has shown that the natural luteinizing hormone releasing factor (LHRF) produced in the hypothalamus is comprised of 10 amino acids arranged in a specific sequence and conformation. The glutamic acid is arranged in a ring form (pyro-), and there is an amino group (NH$_2$) at the other terminus of the molecule.

a complication because LHRF also stimulates FSH release to a certain extent and, if this is prevented, then the entire ovarian set of follicles might be affected to some degree. Another theoretical possibility would be to prepare antiserum against synthetic LHRF and administer the antiserum as a contraceptive treatment that prevented LH release and hence inhibited ovulation. But once again these treatments could affect the entire reproductive system, including important effects about which we know almost nothing now. Further study obviously is needed in this area of our knowledge.

Prostaglandins might prove useful as contraceptives since injection of these fatty acid derivatives into monkeys that have just been mated causes a sharp reduction in the secretion of progesterone by the corpus luteum, thus leading to expulsion of the embryo two days after treatment. This occurs because the uterine lining is not receptive to implantation when the progesterone level is low. A prostaglandin thus would constitute a postcoital contraceptive.

In addition to other lines of research, some particular types of oral contraceptives would be very useful if they can be found and developed. For women, a contraceptive that would be taken only once per month to make the uterus unreceptive to implantation would be desirable as long as the effect was restricted to this organ. A preovulation test might be found, so that a woman would abstain from sexual intercourse for the three days before she knew the ovum would be released (two days for the lifetime of the sperm and one day for the lifetime of the egg). Indications of an oncoming ovulation might be derived from some test that showed that a follicle was about to ripen, or that some substance was excreted in the urine or saliva a few days before ovulation. Some progress actually is being made in these lines of study. Postcoital contraceptives could be used to cause expulsion of the fertilized egg, even without knowing whether or not conception had occurred. Some chemicals have such an effect, and these have been used successfully in some cases of rape. But too little is known about the toxicity or side effects of most of these substances at the present time.

Some efforts have been directed toward the development of male contraceptives, but none has been marketed because of side effects or some observed toxicity. It is curious that the obvious side effects or even lethal effects of oral contraceptives for women have not deterred their exploitation by most of the medical and pharmaceutical professions. Some of the rejected male oral contraceptives have included chemicals that arrest spermatogenesis, but these are reported to produce toxic effects or hyptertension or other undesirable symptoms. One of these agents was reported to stop sperm production and to have no toxic effects, but it led to the development of a severe hangover if it was taken before drinking even a small amount of alcohol. Further investigation of this effective antifertility agent apparently has been abandoned because of its judged impracticality for marketing. The dual standards that operate for men and women in our society seem to have had some influence in this situation, and some day such decisions may be redirected toward fairness to both sexes.

With improving technology and a better understanding of biological phenomena, a number of alternative approaches have been investigated with the view of obtaining long-term methods for contraception. Among these are long-term ("20-year") pills containing hormone derivatives that would be implanted under the skin in women, and vaccination that could render a women immune to sperm and thus prevent fertilization by incapacitating any sperm introduced during coitus.

Sterilization

Surgical prevention of parenthood can be a permanent means for achieving fertility control. Surgical **sterilization** is virtually one hundred percent effective in preventing pregnancy. In women, the two Fallopian tubes, or oviducts, are cut and then tied off so that ova cannot reach the uterus nor can sperm reach the egg across the severed passageway. In men a **vasectomy** is performed by cutting the vas deferens from each testis and tying off the ends, thus preventing sperm in the vas deferens from moving into the ejaculatory ducts (see Fig. 5.5). Neither procedure involves removal of the sex glands or sex organs so that the hormonal physiology remains intact, and sexual desire and performance are *not* affected. Women continue to ovulate and undergo menstrual cycles, but the ovum remains in the blocked oviduct and eventually degenerates. Its resorption by the surrounding tissues is a natural occurrence. Men continue to produce seminal fluids and to ejaculate these during orgasm, but no sperm are present. Sperm continue to be produced in the testes but they cannot pass through the blocked vas deferens, and these gametes also degenerate and are resorbed, just as happens every day when a man fails to ejaculate even with an intact passageway. The vasectomy is a simple operation in which small cuts are made on both sides of the scrotum, and the vas deferens then are cut and tied. No hospitalization is required. For women, on the other hand, an abdominal incision must be made to provide access to the oviducts, which then are cut and tied off; a few days of hospitalization usually follow the operation.

Recently improved methods permit the restoration of an intact vas deferens in a few cases. The duct ends are sewn together, and the physician can perform the precise and delicate maneuvers more confidently since magnification is used. Special talents are required to perform the reverse operation successfully, so the success rate is still very low. A similar operation to reunite the cut ends of the Fallopian tubes also is possible, with similar restrictions on success as for a vasectomy.

Sterilization is legally permissible in the United States and many other countries, although medical reasons rather than personal choice may be required before the operation can be performed. There has been a vigorous program in India to encourage men to undergo vasectomy. The program is subsidized, and about one million men have taken advantage of the opportunity during the past decade.

Abortion

Abortion involves the spontaneous loss (*miscarriage*), the induced loss, or the removal of the growing embryo or fetus from the uterus. Miscarriages may occur at any time but are most frequent during the second and third months of pregnancy. Many different causes may lead to spontaneous abortion, originating in either the mother or the fetus. The estimate is that at least 10 percent of the pregnancies in the United States terminate prematurely for one reason or another leading to spontaneous loss, but this figure is much higher when induced abortions are included in the estimates.

One of the commoner causes for spontaneous abortion is the occurrence of an unbalanced chromosome set in the fetus. About 40 percent of the spontaneously aborted fetuses that have been studied have been found to have one extra chromosome. There is an interesting correlation between the time of miscarriage and the kind of extra chromosome present. The earlier the abortion, the longer the extra chromosome tends to be; and the extra chromosome tends to be one that is shorter and shorter as the abortion occurs at later and later times. One exception that is known is for chromosome-22, one of the shortest in the entire human complement (see Fig. 3.1). This chromosome tends to be found in very early cases of abortion, probably because of unbalances that are crucial during earliest embryonic development. Although chromosome-21 is found in a substantial number of aborted fetuses, the difficulties due to the presence of an extra chromosome-21 are not always sufficient to cause death of the fetus. Infants with an extra chromosome-21 develop Down's syndrome, or mongolism, and this is one of the most frequent of the chromosome anomalies to survive until birth and afterward.

Loss or addition of the **X** chromosome clearly leads to physical, mental, and physiological deficiencies in males and females, but not to the same high frequencies of miscarriage as have been found for autosomal unbalances. The variable effects of different chromosome unbalances most probably depend on the critical genes required for normal embryonic development. In the less severe cases of chromosome unbalance, the baby is born but may show some affliction that is due directly to the chromosomal anomaly.

Therapeutic or induced abortions generally are performed no later than the sixteenth week of pregnancy. In many countries the practice of abortion is illegal, especially in regions that are predominantly Roman Catholic. In the United States and other countries that legally permit abortion, whether by local statute or national law, the abortion may be obtained by choice of the woman or only after the recommendation has been made by a physician if the woman's life is in danger or if there is a high risk of abnormality in the fetus. Abortion is legally available in Japan to any woman who wishes one, and about one million abortions per year are performed in that country. In Hungary, where abortions are free, there are more abortions than live births each year. Where abortions are illegal there still may be a high rate of induced termination of pregnancy before live birth, but the incidence of physical damage to the mother, or even death, is much higher. In such situations the abortions either are performed by poorly trained practitioners or under conditions of minimal sanitation and medical care. Removal of the embryo or fetus from its uterine lining involves traumatic and physically damaging procedures. If this is not performed properly under suitable and clean conditions, then the woman may suffer extensive hemorrhage; uterine tissue damage that may be irreparable and lead to subsequent sterility; ill health or death from infectious disease contracted in the course of the operation; or any number of problems ordinarily not encountered in the operating room of a legally sanctioned physician's office or hospital.

The arguments concerning induced abortion are varied and complex. To most people it is reprehensible to terminate a life; but life is defined in different ways by different people. The fertilized egg is a living cell, and an embryo or fetus is made up of many living cells. But whether these are living cells like any others in the body or whether these cells are to be viewed as a special life, separate from the mother's, is the crux of one of the controversies. There are no absolute answers to the serious questions that have been raised, and there probably will continue to be varying opinions because of differences in religious, ethical, social, and personal persuasions.

It is a fact that the introduction of legalized abortion in New York State led to a dramatic decrease in deaths from abortion and to a significant decline in the birth rate since the law was introduced in 1970. These favorable statistics did not occur in other states that viewed abortion as criminal; states where persuasive and explicit medical reasons had to be presented to a physician, a group of physicians, or a team of legal and medical personnel who then made the decision as to whether or not a requested abortion was to be performed.

Infertility

The inability to procreate may be total (**sterility**) or temporary (**infertility**). It has been estimated that as many as 25 percent of the couples who wish children either have none at all or fewer than they would like and plan to have. For those couples who are fertile, physicians have estimated that about 50 percent accomplish conception within one month of sexual intercourse; most achieve successful pregnancy within six months. The causes of infertility or sterility are varied. About half of the cases result from some factor in the woman, about 30 percent are due to the male, and the remainder of the incidents of childlessness are due to a problem in both partners.

Causes of Childlessness

The general causes of infertility or sterility are physical defects, emotional stresses, ovulation or implantation problems, and inadequate functional sperm counts. Sexual performance usually is unaffected by the underlying hormonal or physical problems that lead to a deficiency or lack of viable gametes, but emotional and psychological factors that are derived from continued failure to achieve a desired pregnancy certainly could interfere with the pleasure and activities of the sexual relationship.

The chief problem in men who are infertile is an inadequate production of viable sperm. Although the estimates vary, it is generally assumed that a minimum count of 10-20 million normal sperm per milliliter of semen is required for success in fertilizing the egg. Although the actual sperm count may be much higher, infertile men may have an excess of inviable gametes in the total volume of semen and thus be unable to initiate conception. Low sperm count, an excess of infertile sperm, or lack of sperm may result from any one or more causes. Inadequate nutrition, poor health, emotional stress, and physical problems such as the blockage of the sperm ducts usually are reversible and can be corrected by suitable treatment. Some causes of infertility in men may not be correctable, such as overexposure to irradiation that leads to abnormal spermatogenesis, or some birth defects, or diseases such as mumps that also may interfere with sperm production if the illness occurred during adulthood.

Because of the more complex reproductive anatomy and physiology of women, the causes of failure to conceive are more numerous. In some cases the oviduct or uterus may be blocked or deformed, and corrective surgery may be possible. Some women develop an allergy to sperm proteins and reject sperm because of immunological incompatibilities. Such conditions often are temporary, and the allergy may disappear if the man uses a condom for a time when intercourse is performed or if both partners abstain from coitus for an empirically determined interval of time. The main reasons for the woman's inability to become pregnant are related to problems of ovulation and implantation of a fertilized egg. Failure to ovulate or to implant the embryo usually is correctable by hormonal treatment or by appropriately prescribed medication. Examination of uterine tissue specimens may reveal a problem with implantation, while indications of ovulation failure sometimes can be obtained from daily readings of the basal body temperature. Although not completely reliable, many women show a slight increase in body temperature about 10 days after menstruation has ended. The lack of a brief, periodic temperature rise might indicate that ovulation was not taking place, at least for some women.

Emotional stresses often lead to temporary infertility, in both men and women, perhaps because of the participation of the central nervous system in the timing and production of pituitary hormones under the guidance of the control center of the hypothalamus in the base of the brain. The interactions among hypothalamic, pituitary, and gonadal activities are intricate and precise. Unbalance in any one or more of these systems leads to some level of dysfunction in the reproductive physiology of the individual. Studies have shown that men under stress (college students at examination time, soldiers in combat or preparing to enter an area of danger, and individuals near physical or emotional exhaustion from various causes) all have markedly lowered sperm counts at these times. Despite these observations of the influence of emotional stresses on reproductive capacity, women who have been subjected to the trauma of rape may become pregnant while others who want a child may be unable to conceive. These areas of our knowledge are filled with unknowns.

Remedies for Childlessness

For couples who cannot have children for whatever reason, **adoption** can start or complete a family. Although not in widespread use, **artificial insemination** is available in various countries, including the United States. In this procedure the physician uses some mechanical means to introduce semen from a donor into the reproductive tract of a woman. If conception occurs, then a normal pregnancy usually ensues and the baby has received half its genes from its mother and the remaining half from the sperm donor. Although still in the experimental stage, **embryo transplantation** has been attempted in humans. Laboratory investigations using mammals such as mice and rabbits have provided a substantial body of information useful in initiating the possibilities of transplanting a very small embryo into the receptive uterine lining of a prospective human mother. Problems of immunological incompatibilities, as well as our vast ignorance of human reproductive physiology, must be overcome before this method becomes practical. Similar problems confront the medical profession in accomplishing successful transplants of vital organs such as the heart, kidney, lungs, and others.

Although some women have been able to conceive after initiating a program of taking oral contraceptives, this is not the usually recommended procedure. For the women who have had success (in one study it was shown that 69 out of 85 women conceived within two months of discontinuing the pill), the effect apparently is similar to the "rebound" that has been experienced by normally fertile women after they have discontinued a regime of oral contraceptives. The usual fertility drugs contain FSH, FSH-like, or LH substances that are isolated from various sources, one of which is the urine of postmenopausal women who have a particularly high concentration of gonadotrophic hormones. Some infertile women who take the FSH fertility drugs do exhibit induced ovulation. But in about 50 percent of these women there have been multiple births, with as many as seven babies in a single pregnancy. Many of these multiple births occur prematurely.

New kinds of fertility drugs have become possible because of the recently developed method of synthesizing the LH-releasing factor. The synthetic chemical does stimulate the secretion of both LH and FSH in women who do not ovulate but whose ovaries and pituitary gland are functional. In the cases of infertility that are not due to a hypothalamic-pituitary defect, injections of synthetic LH-releasing factor can stimulate ovulation. There is a greater probability that a normal pregnancy will be initiated after FSH and LH secretion have been induced by the injected chemical. Infertility in males that is based on inadequate secretion of gonadotrophic hormones also could be alleviated by such therapy.

Fertility clinics exist in many regions of the United States and other countries. Medical and counseling services should be sought by couples who wish to have children but have not been able to achieve this goal. Similarly, information about fertility or antifertility can be obtained from local Planned Parenthood Committees and other agencies with regional or national headquarters. Abortion referral should be sought from established medical groups or individuals, and not from sleazy advertisements in newspapers and magazines. The dishonesty or greed of some recently established abortion agencies, and the irreparable damage that could be inflicted by irresponsible treatment, argue against a shoddy approach to this important therapeutic possibility.

Suggested Readings

Berelson, B., and R. Freedman, A study in fertility control. *Scientific American, 210* (May 1964), 29.

Blake, J., Abortion and public opinion: The 1960–1970 decade. *Science, 171* (1971), 540.

Carter, L. J., Contraceptive technology: Advances needed in fundamental research. *Science, 168* (1970), 805.

Carter, L. J., New feminism: Potent force in birth-control policy. *Science, 167* (1970), 1234.

Chang, M. C., Mammalian sperm, eggs, and control of fertility. *Perspectives Biol. Med., 11* (1968), 376.

Djerassi, C., Steroid oral contraceptives. *Science, 151* (1966), 1055.

Edwards, R. G., Mammalian eggs in the laboratory. *Scientific American, 215* (August 1966), 72.

Edwards, R. G., and R. E. Fowler, Human embryos in the laboratory. *Scientific American, 223* (December 1970), 44.

Glass, R. H., and N. G. Kase, *Woman's Choice: A Guide to Contraception, Fertility, Abortion, and Menopause.* New York: Basic Books, 1970.

Guillemin, R., and R. Burgus, The hormones of the hypothalamus. *Scientific American, 227* (November 1972), 24.

Neel, J. V., Some genetic aspects of therapeutic abortion. *Perspectives Biol. Med., 11* (1968), 129.

Oster, G., Conception and contraception. *Natural History, 81* (August-September 1972), 46.

Pincus, G., Control of conception by hormonal steroids. *Science, 153* (1966), 493.

Stern, E., Contraceptives and dysplasia: Higher rate for pill choosers. *Science, 169* (1970), 497.

Stone, A., The control of fertility. *Scientific American, 190* (April 1954), 31.

Tietze, C., and S. Lewit, Abortion. *Scientific American, 220* (January 1969), 21.

part three

SOCIAL AND REPRODUCTIVE BEHAVIORS

chapter eight

General Aspects of Animal Behavior

Features of the System

By **behavior** we usually mean adjustive responses to environmental variation. The response may lead to an internal or external adjustment in the organism, and the environmental factors may be physical or biological components of the surroundings. Behaviors usually involve a coordinated set of muscular activities in animals, rather than exclusive reference to what an animal does. Some characteristics of what an animal does, such as growth, would be outside the concept of behavior that thus calls for a more circumscribed view of the phenomena. Muscular activities may be localized, as when a dog wags its tail or a person swats a fly; or they may involve more of the anatomy, as when an individual runs or swims. Some behaviors lead to a cessation of movement, such as the possum that plays "dead" or the pointer that "freezes" at the sight of a game bird or a soldier who stops moving when confronted by an undetermined sound or sight.

Changes in the internal environment also may initiate behaviors. One of the better known situations occurs when a mammal experiences thirst. This condition is triggered by a deficit of water in the blood that stimulates the **hypothalamus** at the base of the brain to signal the release of the antidiuretic hormone **vasopressin** from the posterior lobe of the **pitui-**

tary gland, which causes the kidneys to resorb more water. The animal also becomes restless and searches for water. Once the thirst has been satisfied and internal neurochemical changes have taken place, the animal becomes less restless or quite peaceful and may begin to groom, or go to sleep, or read a book.

All animals encounter problems due to the variable nature of the environment. The survival of the species requires that the animal live long enough to reproduce. Thus it must avoid or overcome inhospitable external factors; secure raw materials and energy for metabolism; and find a suitable and appropriate mate, achieve successful mating, and in many cases guard and educate its young. Diverse behaviors accomplish these ends, and no two species achieve these successes in precisely the same way. Similar situations evoke different behavioral responses in different species, and different situations elicit different and appropriate behaviors from individual animals. Both phenomena reflect different patterning of muscular contractions. In the same way, different behaviors of an individual from one moment to the next reflect different patterns of muscular activity.

Adaptive, meaningful responses to particular environmental variables require systems of **receptors** (sense organs) that detect changes, integrated nervous and hormonal (**neuroendocrine**) systems through which the information is received and the response is triggered, and systems of **effectors** that are organs that carry out the responses (Fig. 8.1). Some animals release noxious substances in response to particular stimuli, or produce electricity, or change color, among other things. Humans and other mammals possess muscles and glands as primary effector systems; receive information about their internal and external environments through receptors such as ears, eyes, nose, nerve endings in the skin, muscles, and other systems; and coordinate these sensory inputs via nervous and endocrine gland systems. When an infant suckles at its mother's breast, the hypothalamus causes oxytocin to be released from the posterior pituitary in response to this stimulus, and this in turn stimulates muscular contractions around the milk ducts in the breast such that milk is forced out into the ducts leading to the nipple of the breast (see Fig. 6.10).

Since behavior usually is some movement, lack of movement, or secretion in response to an external or internal environmental change, it depends on effector systems and on the capacities and capabilities of the nervous system. The role of the environmental variable that acts as the stimulus depends on how the receptor, effector, and neuroendocrine systems are integrated and interconnected. The organization of these three systems varies among animal groups and even within the same individual relative to the specific stimuli presented. In some cases the stimulus triggers a simple set of muscular actions that proceed auto-

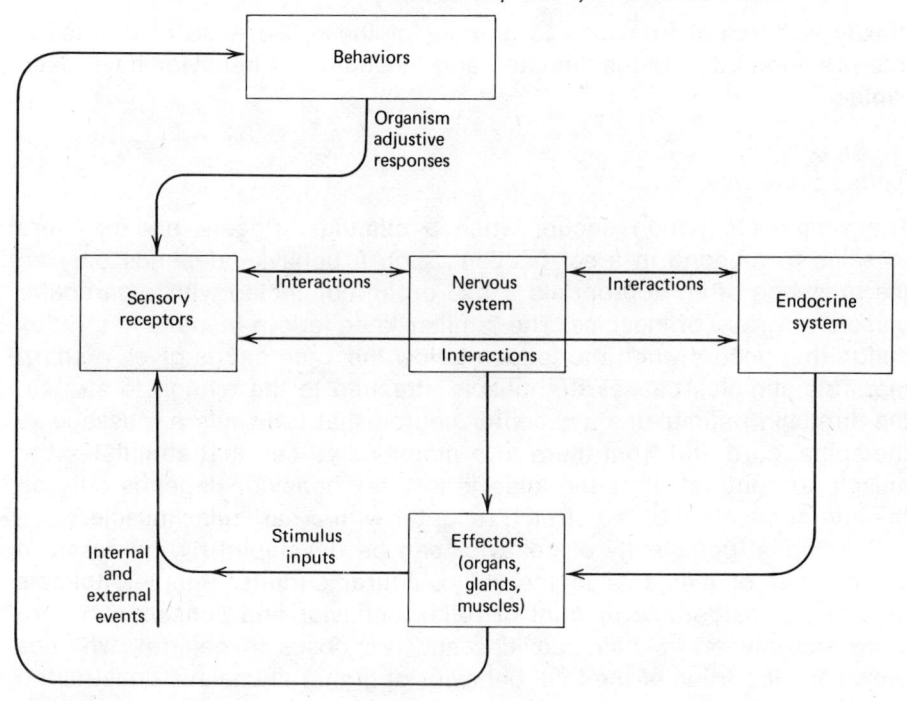

figure 8.1
Schematic diagram indicating the systems involved in unitary control governing behaviors.

matically once started; or, as the muscular action proceeds it may be monitored constantly by stimuli that it has generated in the body and thus direct the action; or, the stimulus may trigger a complex response that is centrally patterned in the nervous system.

Although it is convenient to categorize modes of behavior, there is a continuum of expressions throughout these sequences of difference. No animal behaves exclusively according to one behavioral mode, but the *predominance* of one or more modes over the other capabilities usually can be recognized. Human beings exhibit diverse behavioral modes, but the predominant theme in human behavior is based upon learning rather than on more stereotyped patterns.

Stereotyped Behaviors

In many animals, whether simple or complex, a particular stimulus may elicit a predictable and unvarying response. There are a number of behaviors that conform to this pattern, but the biological basis and complexity is different from one to another of these. Because of erroneous interpretations, the terms "innate" and "instinctual" behavior have been avoided.

Reflex Behavior

The simplest situation occurs when a stimulus triggers one or more muscles to respond in a set fashion. Such a behavior depends only on the presence of an appropriate sense organ connected with a particular muscle or group of muscles. The familiar knee jerk in humans is a **reflex** action that occurs when the tendon below the knee cap is given a sharp tap. This stimulus causes the muscle attached to the tendon to stretch; the stretching stimulates a receptor neuron that transmits a message to the spinal cord and from there to a motor nerve cell that stimulates the muscle to contract; thus the knee jerk reflex behavior depends only on the interconnection of the stretch receptor with a particular muscle.

Substantial complexity of behavior can be developed by increases in the number of elements in the reflex neural circuits. Simpler animals display a considerable amount of reflex behavior and consequently are more stereotyped in their activities and responses in general, whereas only a small portion of the total behavior of higher animals is constructed of reflexes (Fig. 8.2). Interconnected reflex circuits provide one means for some remarkably regulated and coordinated complex behavior in insects, and they also underlie components of sexual behavior in mammals and other vertebrates.

Erection of the penis and ejaculation of semen basically represent spinal reflex processes in the human male. The principal sensory components are highly sensitive mechanoreceptors located in the tip of the penis that initiate impulses that are transmitted via nerve fibers to the lower spinal cord. Neural signals are then sent to the arterioles of the penis erectile tissues, which lead to vascular engorgement and penile erection and to glandular secretions that are added to the seminal fluids. Ejaculation occurs coordinately by a similar or identical spinal reflex system. Once the neuroendocrine systems have received the informational inputs, the responses are elicited from the smooth muscle of the internal reproductive ducts and the skeletal muscle at the base of the penis. The overall response occurs in two phases: (1) the genital ducts contract, thus emptying their contents into the urethra; and (2) the semen then is expelled from the penis by a series of rapid muscle contractions. The rhythmical contractions of the penis are maintained by a pattern of automatic nervous discharge during ejaculation. In addition to these principal events, other muscular activities also are initiated by the same reflex pathways. In humans, however, the higher brain centers may exert

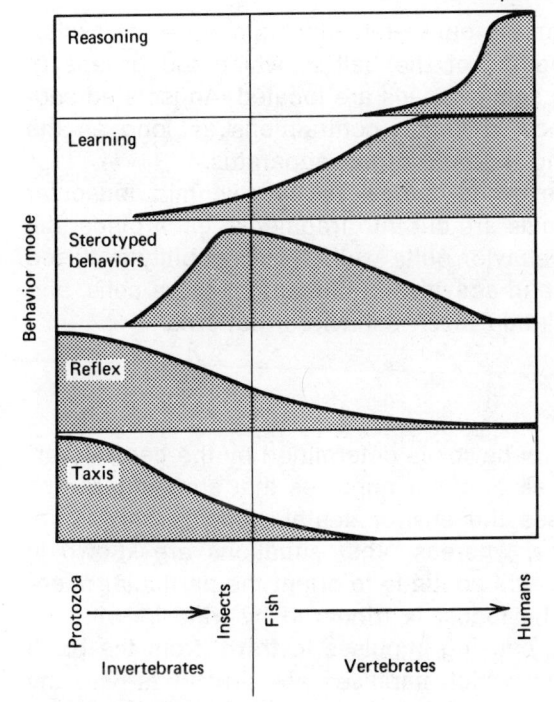

figure 8.2
The dominant modes of behavior according to evolutionary development of animal groups. (Modified from V. G. Dethier and E. Stellar, *Animal Behavior.* Englewood Cliffs, N.J.: Prentice-Hall, Inc., 1961. Reprinted with permission.)

stimulatory or inhibitory control over the reflex behaviors by their regulation of nervous pathways. Erection thus may be facilitated in the absence of penile stimulation and, conversely, erection may be inhibited by psychological factors even though appropriate sensory inputs are made.

Spontaneous Activities

Some types of spontaneous behavior occur independently of external stimuli and are a principal component of the total activities of simple animals that possess recognizable groups of nerve cells. In cases of **spontaneous activity**, the same performance can be accomplished by the whole individual as well as by a small portion of the animal, so long as an intact element of the nervous mechanism is retained in that fragment.

Jellyfish manage smooth swimming movements by virtue of the rhythmic contractions of the bell forcing out water. Rhythmic waves of excitation originate in marginal regions of the bell in which the groups of neurons and gravity-detecting sense organs are located. An isolated segment of the bell still will show rhythmic contractions as long as the segment contains neurons and a sense organ apparatus.

Various worms exhibit burrowing behaviors by rhythmic muscular contractions. When such animals are cut into fragments, each piece still may be able to perform the behavior quite well. This capability is based on the intrinsic programming and activities of clusters of nerve cells, and a brain is not required to maintain these behavioral patterns.

Fixed-Action Patterns

The **fixed-action** patterning of behavior is determined by the central nervous system through sequences of motor impulses that are triggered by specific stimuli. In some cases the environmental stimulus only is required to initiate the behavior, whereas other situations are known in which the environmental clues will continue to orient the particular fixed-action pattern. The flight of the locust is triggered by wind blowing on special receptors in the head, causing impulses to travel from the brain to the region of the body from which impulses are sent in turn to the motor nerves responsible for wing muscle movements. Flight thus is a consequence of a complicated patterning of muscle activities, the sequence being determined by the nervous system. The stimulus of wind initiates the pattern, but its precise sequence and characteristics are programmed in the central nervous system of the insect.

Many ground-nesting birds demonstrate an egg-rolling response which is continued so long as the egg continues to wobble from side to side while being rolled back to the nest. The fixed-action pattern of retrieving the egg is triggered by the sight of the egg outside of the nest, and it is evidenced by the bird moving its bill away from and toward its breast in repeated movements. At the same time the bird usually rolls the egg back to the nest using side-to-side motions that presumably prevent the egg from slipping away. The forward-and-back motion is elicited by the sight of the egg, but it will be continued even if the egg is removed after the behavior has started. This represents a fixed-action pattern. But the side-to-side movement will stop if the egg is replaced by a cylinder. In other words, the side-to-side motion is maintained as long as stimuli come from a wobbly egg, but this portion of the egg retrieval pattern is independent of the beak forward-then-back motion that will continue, once triggered by the sight of the egg outside of the nest, even when the egg later has been removed.

Taxis and Tropism

The automatic orientation of the body in relation to some particular environmental factor is called a **taxis** when active locomotion is involved, as in most animals and free-moving organisms; it is known as a **tropism** if the response consists of growth, as in most plants. The response may be positive or negative, depending on whether the orientation is toward or away from the stimulus, respectively. The stimulus generally is detected in some receptor cells or organs and is transmitted to hormonal or neuroendocrine systems that provide the materials by which the effector organs or cells may carry out the response. The gathering of insects around a light comprises an example of positive *phototaxis* and superficially resembles the same phenomenon as the movement of unicellular algae or protozoa toward light. The anatomical complexity that underlies the insect behavior is of much greater magnitude and intricacy, but it still can be categorized as a taxic behavior.

The common observation that plants bend or grow toward the light can be explained as a tropic response. More growth hormone is secreted in the cells farthest from the light source that causes that side of the stem to grow faster than the more exposed region. The relatively greater growth on the one side leads to a "bending" effect. If plants are rotated, as those on a window sill, then the averaging effect is straighter growth. Plants in their normal habitat are presented with light coming from several directions as the sun moves from east to west each day, which produces a straight growth habit as an overall effect.

Various environmental factors will elicit taxic and tropic responses, including chemicals, water, sound frequencies, and others. Taxic behavior comprises a substantial component of the activities of most lower organisms. A taxic behavior in moths forms the basis for directing males specifically to females of their own species, even miles away. Females produce **pheromones**, or sex attractant chemicals, which stimulate the males to fly to their source and, once there, they accomplish mating by which the species is perpetuated.

Motivated Behaviors

Behaviors that are associated with some *deprivation* are called **motivated**, and these patterns are highly specific to particular stimuli and the particular state of the individual. Some examples of such behaviors are drinking, feeding, courting, copulating, nest building, defense of territory, and care of the young. Motivated behaviors in mammals principally are influenced by the forebrain, particularly the hypothalamus and the cerebral cortex (Fig. 8.3).

Neural Influences

The hypothalamus is considered to be the control center for much of motivated behavior and exerts a substantial influence in relation to hunger; thirst; sleep; emotional, sexual, and parental behaviors; and others. Depending on which hypothalamic area is stimulated, destroyed, or manipulated experimentally, a rat may exhibit profound changes in its eating pattern. Destruction of one region may cause the animal to eat prodigiously and double its weight, although damage to a different region will lead to starvation even in the presence of ample food. Thus there are inhibitory and excitatory centers in the hypothalamus. Electrical stimulation may cause the rat to continue to eat even when satiated or to stop eating even if it is hungry, so that mere damage to the hypothalamus is not the basis for modified behavior. Specific regions exert specific levels of control over eating behavior, among other patterns. Placement of sex hormones into specific areas of the hypothalamus elicits mating behavior; warming of implanted wires causes the rat to pant while the cooling of these same wires in the brain will cause the animal to shiver. If available, the rat will press a bar to turn on a heat lamp or to receive a blast of cool

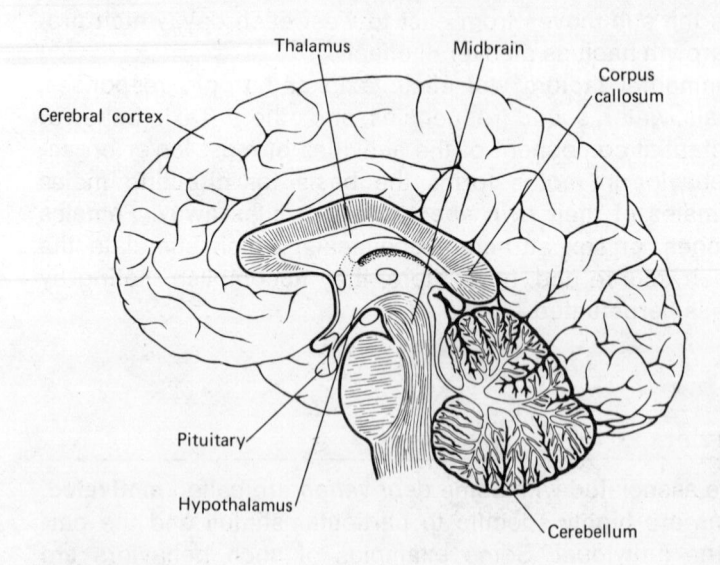

figure 8.3
Diagram showing regions of the human brain. Note the extensive development of the cerebral cortex that essentially overlies the remainder of the brain.

air in response to temperature changes of the wires implanted in its hypothalamus.

The forebrain has enlarged considerably during vertebrate evolution, and the cerebrum has developed into the most prominent portion of the forebrain, reaching its greatest absolute and relative size in humans. The inner "old" cortex and the outermost neocortex (*neo*, new) regulate the conscious, emotional, and intellectual capabilities of the individual human being and other mammals. The hypothalamus, old cortex, and associated structures play a substantial role in governing emotional behaviors such as fear, rage, sleep, feeding, sexual activities, and others; whereas the neocortex appears to modulate these emotional displays by an overriding rational behavior potential. The neocortex exercises a restraining effect on the functioning of the older *limbic system* (hypothalamus, old cortex, and other structures) and its mediation of basic emotions. Most of the evidence in support of the relative functioning of the cortical regions of the cerebrum has been obtained from observations of animals that have undergone varying levels of brain damage.

Experiments using the three-spined stickleback have provided numerous insights into the complex of factors that influence and modulate motivational behavior. This fish species demonstrates a complex pattern of reproductive behavior in which the male and female actions provide an alternating series of stimuli and triggered responses (Fig. 8.4). The male migrates, selects a territory, builds a nest, and proceeds to court any gravid (egg-bearing) female who enters his territory and swims in a particular fashion. He aggressively prevents any other kind of stickleback from entering his territory. The male courts a female by performing a "zigzag dance," and the female follows him. He leads her toward the nest and, as they arrive, the male points his head toward the entrance. The female is stimulated to enter the nest in response to this pointing; he in turn is stimulated to prod her flanks, known as "trembling;" she is stimulated to spawn by this trembling. The male then fertilizes the eggs and further sexual behavior is terminated by stimuli coming from the eggs. At this point the male resumes his previously aggressive behavior pattern and begins to care for the nest.

These motivational levels have been shown to be related to changes in state in the central nervous system that are evoked by such nonneural factors as hormones, blood composition, and blood temperature. Specific external stimuli, such as the assumption of certain postures by the female and the male, trigger or at least influence motivational behavior. A feedback signaling indicates the satisfaction of a particular need, and the next step in the sequence is undertaken, or a new behavior is initiated upon the completion of some previous pattern and the introduction of a new set of stimuli.

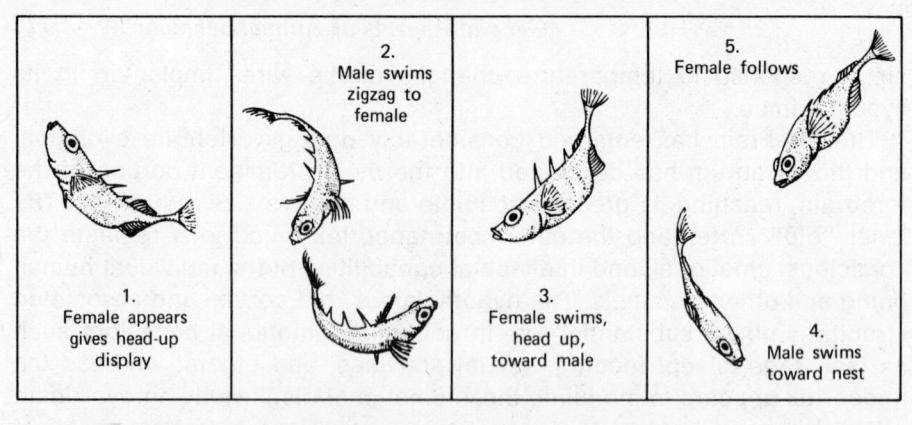

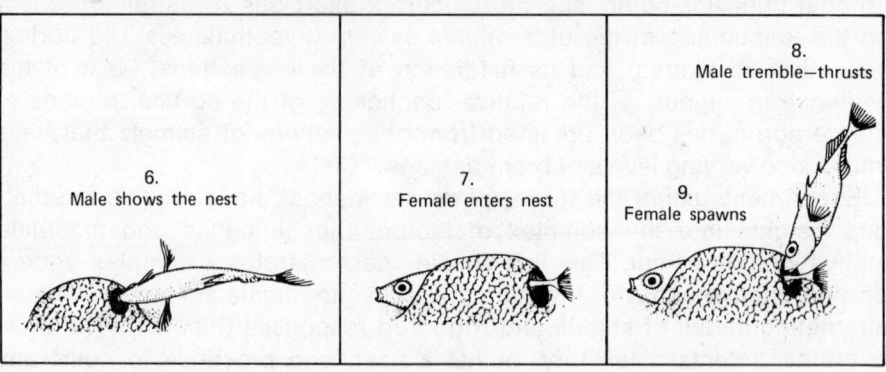

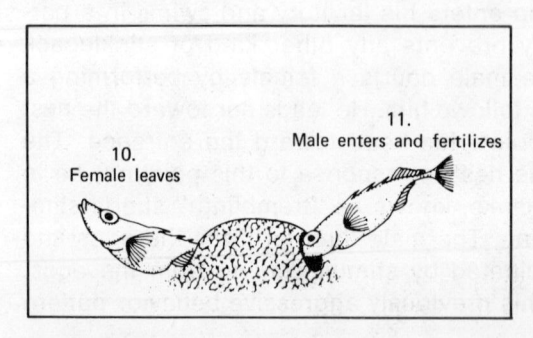

figure 8.4
Courtship and reproductive behaviors in the three-spined stickleback.
(Modified from N. Tinbergen, *The Study of Instinct*. Oxford University Press, 1951. Reprinted with permission.)

Pattern Periodicities

The behaviors thus far mentioned have been those in which fluctuations occurred for reasons that could be experimentally established and elucidated. Some behavior patterns exhibit poorly understood sequences of cyclic fluctuation, which may adhere to daily, monthly, or seasonal schedules. Such fluctuations are associated with specific environmental occurrences, but they may be maintained even in the absence of such factors once the behavior pattern has been set. Daily (**circadial**) cycles are extremely common. Among the primates, the more ancient prosimians exhibit **nocturnal** rhythms whereas the anthropoids, including humans, show a **diurnal** activity pattern in being more active by day and quiescent at night. Various animals show a lunar cycle of activities that may be of startling accuracy. The Pacific grunion of the western United States is a small fish that swarms on the three or four nights when a spring tide occurs during April, May, and June. The fish squirms onto the beach at precisely the high point of the tide when sand deposition is greatest, the female wriggles into the sand tail-first and deposits eggs, the male curls around her body and ejaculates sperm that fertilize the eggs, and the grunions rush back into the ocean at the next wave. By the time the next tide reaches the spots on the beach where the fish had left the fertilized eggs some 15 days earlier, the young fish have hatched and are ready for their own independent life in the sea.

The estrous or menstrual cycles of most mammals occur at fairly regular intervals, varying from a few days to as long as one year in animals with a single breeding season. The tempo of the biological clock is set by an internal rhythm although the clock itself often is set by some environmental stimulus. Rhythmic behavior continues even when the animals are placed in a constant environment, as when nocturnal species are kept in constant daylight. They still keep track of the nighttime hours. Cycles can be reset under new environmental conditions, too. If spring-breeding animals are transported to another hemisphere, their cycle eventually shifts to coincide with spring time in their new home. Humans who suffer initially from "jet lag" reset their biological clock after a few days to a week, depending on the number of hours difference between the point of origin and the final destination. If the usual waking hour in New York is eight a.m., then the traveler to California may awake at five a.m. for several days, but soon becomes adjusted and rises at a more reasonable hour if the visit lasts long enough. Travelers going long distances usually are advised to stop over at intervals along the way so that their biological clock can readjust at an appropriate rate.

Hormonal Influences

As the individual matures, the endocrine glands also grow and develop. The hormones secreted by these glands affect behavior by altering the internal environment in which the nervous system operates. Such alterations of behavior may include effects due to the enhancement of organ development involved in behavior, effects on the early development of the nervous system and on sensory inputs to the neural apparatus, influences on special centers in the central nervous system, and through various nonspecific effects on the animal as a whole.

The development of secondary sex characteristics in humans and other animals occurs under hormonal influences. The horns and antlers of many male mammals are important in courtship, aggressive behavior, and mating success; and these structures develop under hormonal influences as the animals become sexually mature. Neural changes induced by hormonal action often are associated with the greatly increased rates of growth of the organs upon which these same hormones act specifically. The growth of the uterus, mammary glands, and other organs in female mammals is influenced by the steroid hormones estradiol and progesterone.

Nest building by rats is known to be enhanced by lowered body temperature, and reduction in temperature results from the removal of glands such as the thyroid. In these animals there is a general lowering of body temperature and an enhanced activity in nest building. This illustrates the effect on behavior of hormones that influence the general body conditions in the animal as a whole.

Hormones generally influence behavior by direct action on specific areas of the central nervous system. The effect occurs primarily on the hypothalamus in mammals. Experiments with cats have shown that a spayed female may undergo morphological changes of reproductive organs and tissues if given small doses of natural estrogen through the bloodstream over a period of time. Such cats, however, do not display the usual behavior actions shown by normal female cats in heat. If solid *stilbestrol dibutyrate*, a synthetic estrogen that mimics the action of estradiol, is implanted directly into the hypothalamus of a spayed cat then she will develop the expected sexual behavior pattern despite having an unprepared reproductive tract, that is, tissues in the *anestrus* state. Ovariectomized rats also will display sexual behavior if estradiol is implanted directly into specific regions of the hypothalamus.

In general, the role of hormones in the development and expression of behavior has been demonstrated in many vertebrates by a variety of experimental methods. Animals fail to display certain adult or sexual behaviors if they have been castrated when young; premature sexual behavior can be induced by injecting hormones into sexually immature animals; and sexual behavior can be restored by injecting or implanting appropriate hormones into castrated animals.

These hormonal effects apply to recurrent patterns of behavior that occur in animals having cyclic reproductive periods, as well as to the development of behavior during preadult intervals. Hormones play a significant role in the periodic changes in behavior that are related to such reproductive cycles, as we discussed in Chapters five and six.

The extensive studies of the ringdove conducted by Daniel Lehrman showed clearly that hormonal secretions and reproductive behavior were modulated by enivonmental influences. If a female ringdove is isolated from other birds, then her hormonal level remains low and she will not lay eggs. Development of the reproductive tract and subsequent egg laying can be stimulated by the injection of hormones or by permitting her to see a male dove that acts like a male. Upon the introduction of the stimulus of a male dove, the sight activates hypothalamic centers that control pituitary secretions. FSH and LH are released from the pituitary, stimulating growth of the ovary that then secretes estrogen and progesterone; and egg laying then will occur. She will not respond in this manner to a castrated male that displays no sexual activities. A similar pattern was noted for incubation behavior in this species. Unless she is given an injection of progesterone, a female ringdove will not incubate a nest full of eggs so long as she is alone in the cage. If a male is introduced into the cage in lieu of the progesterone treatment, then the two birds will begin courting and, shortly afterward, both the male and female will begin to incubate the eggs. Apparently the courting activities stimulate the production of progesterone, a normal event for animals that live in a sexually mixed society.

Experiential Influences

All the while that the organism undergoes structural and physiological changes that affect behavior it also acquires experience, so that it is often difficult to separate the contributions of the maturation processes from the contributions of experience to the development of behavior. One approach to this problem is to study a particular behavior pattern in hybrids between two species that display variations in the same pattern. In this way, one can determine the genetic carry-over versus the portions of the behavior that are gained through experience by the hybrid animals. Such a study of lovebirds will illustrate this point.

Some species of lovebirds collect nesting material by cutting long, thin strips from leaves or bark (or paper, if provided), but they carry it back to the nest differently. The peach-faced lovebird tucks several strips among the feathers of its rump, whereas Fischer's lovebird uses its beak to carry one strip at a time to its nesting hole in the tree. Hybrids of the F_1 generation act rather confused when building nests because of problems encountered in carrying strips to the nest sites. They retain parts of each parental habit, such that they tuck strips among their feathers, but often in the wrong place; and they may fail to let go of the strips once they have been tucked away. The hybrids become increasingly proficient if they carry strips singly in their bill (the pattern of Fischer's lovebird). But even after two or three years of continuing improvement in perfecting the bill-carrying behavior (the more successful of the two variations in hybrids), the birds still try to tuck the strips among their feathers and never completely become successful carriers so that they remain less successful nest builders. The mode of carrying strips clearly is inherited and not gained by experience, nor can experience overrule the pattern entirely, even when one mode is more successful than the other and either pattern can be employed by a genetic hybrid animal.

Many studies have been difficult to interpret because of the problems encountered in preventing the unborn or unhatched individual from practice and experience while still in the uterus or the egg. Developing embryos are known to behave. The unhatched chick in the egg responds to the warning cries of the rooster by stopping its movements. Prehatch-age mallard ducklings respond vocally to the sounds that they hear from the mother duck as she incubates the eggs. In addition to their making particular sounds at particular times while still in the egg, in response to the mother's calls, it has been suggested that young hatched ducklings may recognize their mother from among other female ducks because they recognize her unique vocalizations, which they heard as they were being incubated. This suggestion remains to be verified in further studies. In general, there is relatively little unequivocal information by which we can decide whether some behavioral pattern is based on genetic patterning, or on experience, or on some combination of the two. These difficulties in interpretations lead to controversies at many levels, including those involved in the questions about genetic versus learned influences on sex roles in human society (see Chapter eleven).

Practice and Experience

Various studies have shown that **sensory-motor coordination** is improved and refined by practice and experience during the development of behavior in higher vertebrates. The same mechanisms are involved in the development of normal performance in the young, the maintenance of normal performance once developed, and adaptation to changes in the environment or way of life. The necessity for practice varies for different species and for different tasks. Thus the chick demonstrates essentially normal distance perception without visual experience and can discriminate patterns even after visual deprivation. Primates, on the other hand, require visual experience for adequate distance perception and for pattern discrimination.

Young kittens find their way to a preferred nipple within the first three days after birth and will continue to secure milk from a particular nipple regardless of the size of the litter. Their experience in finding a particular functional area of their living quarters takes somewhat longer, and depends much more on the maturation of sensory-motor coordination. They become more adept with longer intervals of practice and incorporate their experiences more quickly as they develop physically.

Anyone who has been bedridden for a week or two after some accident or illness knows that it is difficult to walk properly once out of bed because of poor sensory-motor control. Practice rapidly restores the previous expertise to the former patient. The recovery is even more rapid since the child or adult has had previous experience in walking and thus practices the efforts more knowledgeably, leading to a rapid resumption of the learned skill.

Imitation

Learning from other (**imitation**) is quite different from self-learning or practice. Animals that have been raised in isolation have been studied for their development of behavior, uninfluenced by the presence of siblings, peers, and parents as usually would be the case. Although interpretations of the experimental results often are complicated because the isolated animal has been subjected to some sensory deprivations, the great importance of learning has been amply demonstrated in many cases.

Some species of birds develop the full repertory of songs even when raised in isolation, and thus acquire the species-specific sounds by self-learning. Other bird species apparently require some exposure to adults or to hearing their own songs before they can develop a full complement of vocalizations that are characteristic of the species. Both imitation and self-learning are important in the successful development of the singing behavior pattern in birds.

Learning

Learning may be exhibited in various ways and almost certainly has different characteristics and even different mechanisms in animal species. The phenomenon usually is considered to involve processes of incorporating relatively permanent changes in behavior as a consequence of experience, and the retention and recall of these experiences. Learning usually is identified in a negative fashion; that is, as behavioral modification that is not attributable to sensory adaptation, motivational states, endogenous rhythms, and maturation, among other factors. Often it is convenient to analyze experiments according to particular categories of learning, more because of needs in interpreting experimental design than for purposes of elucidating fundamental mechanisms of behavior. These categories are not necessarily recognized by all investigators. More importantly, there is a *spectrum of learning capacities* that may be elicited differently according to the stimuli presented, the mode of reward and punishment for the behavioral responses, and the neural capacity of the individual and the species.

Associative Learning

In general, learning involves the strengthening of responses that aid the animal or are of some degree of importance for its existence and activities. Pavlovian or **classical conditioning** involves a learning by the association of reward and punishment with some previously irrelevant stimulus, as occurred when Pavlov trained dogs to associate the sound of a bell with food. Eventually the dog salivated merely at the sound of the bell even when it could not see, smell, or taste food. The original stimulus of food was replaced by a new stimulus somehow associated with the original, leading to the conditioned reflex. This form of associative learning is one of the simplest to investigate and recognize.

The behavior of the animal controls the situation in **operant conditioning**. A pigeon or a rat presses a lever to obtain some food, and the food is the reward for pressing the lever. The animal determines the frequency of the reward and may even decide how to manipulate the lever to obtain the food, using its head or foot or other structure. Depending on the conditions of the experiment, the animal can modify its behavioral response. If the reward is received only for one kind of action and not for alternatives, the response can be strengthened or **reinforced** by deliberately manipulating the experimental conditions under which the animal behaves. Reinforcement of the response also can be achieved by other means, such as providing an undesirable stimulus that the animal comes to avoid or escape.

Despite the numerous variations due to species differences and experimental designs, some general relationships have been observed in relation to conditioning. The proper *sequencing* of stimuli will elicit associative learning, and *repetition* of trials leads to stronger and more rapid conditioning. Different animals learn at different rates and retain their learned behaviors to varying degrees, depending on intelligence levels and the experimental situations to which they were exposed. Associative learning also may be studied in relation to levels of *generalization* and *discrimination* that are achieved by animals. By generalization, we refer to the ability for response to new stimuli that are similar to the stimulus to which the animal has become conditioned. If the animal can select one particular stimulus from among several that are similar, then it has the capacity for discrimination in associative learning situations.

Reasoning

Reasoning is considered to be the most highly evolved form of learning and it is especially well developed in primates, reaching its peak in humans. Through **reasoning** the animal can form a new experience to achieve a desired purpose when it combines two or more isolated experiences. In a classical example of thinking, a chimpanzee solved the problem of getting a banana that was strung up high out of its reach, by stacking boxes one on top of another and climbing up to get the fruit. This solution to the problem involved abstract reasoning rather than prior specific experience or instruction in achieving this specified goal. In a more abstract situation, some chimpanzees have been taught to associate symbols with particular objects, activities, or desires. Washoe is a chimpanzee who learned about 200 different hand signs that permitted her to communicate with her human companions and to express her wishes. Sarah, another chimpanzee, has learned to create sentences using symbols of different shapes and colors; just as Washoe "spoke" using hand sign language. Symbolic communication represents the highest level of achievement by learning, and it is part of the genetic potential of every member of the human species. But such achievement requires learning to expose the potential for the capacity to communicate realities and abstractions in symbolic forms. Any child can learn any language depending on its social surroundings and its human teachers, so that a particular set or sets of symbols are learned. But in addition to the capacity for such learning, the genetic endowment also must include the physical apparatus by which speech sounds can be formed and uttered meaningfully and understandably. The integration of neural and anatomical systems is required to extract the significant end product that we term intelligence. Intelligence is an outcome of inherited potential that reaches some stage of expression that depends in part on learning and in part on other factors that permit the development and elicitation of the potential that exists.

Suggested Readings

Dilger, W. C., The behavior of lovebirds. *Scientific American, 206* (January 1962), 88.

Hailman, J. P., How an instinct is learned. *Scientific American, 221* (December 1969), 98.

Harlow, H. F., and M. K. Harlow, Learning to think. *Scientific American, 181* (August 1949), 36.

Kandel, E. R., Nerve cells and behavior. *Scientific American, 223* (July 1970), 57.

Lehrman, D. S., The reproductive behavior of ringdoves. *Scientific American, 211* (November 1964), 48.

Pengelley, E. T., and S. J. Asmundson, Annual biological clocks. *Scientific American, 224* (April 1971), 72.

Premack, A. J., and D. Premack, Teaching language to an ape. *Scientific American, 227* (October 1972), 92.

Tinbergen, N., The curious behavior of the stickleback. *Scientific American, 187* (December 1952), 22.

Tinbergen, N., The courtship of animals. *Scientific American, 191* (November 1954), 42.

Tinbergen, N., The evolution of behavior in gulls. *Scientific American, 203* (December 1960), 118.

chapter nine

Sexual and Sex-Related Social Behavior

Although any interactions between at least two individuals may constitute **social behavior,** the usual social relationships involve individuals of the same species. Since sexual and parental behaviors involve at least two individuals, these also may be categorized as social interactions. A social relationship is not necessarily established because individuals aggregate in some localized space, as when moths and other insects cluster around a light, or when many people attend a concert or a cocktail party. Among the invertebrates, many orders of insects are social; and among the vertebrate groups social interactions principally characterize the birds and the mammals. Such animals live in groups of individuals belonging to the same species. The basis for social life fundamentally involves reproductive behaviors, principally the behaviors related to mating and to care of the young. In addition there may be family life, group life, and fighting or other aggressive actions comprising social interactions.

Social Order

Animals that live in groups demonstrate particular social organizations and geographic spacing, or **territoriality**. Although evident in many animal species, such order is especially well developed in the primate groups of monkeys, apes, and humans.

Hierarchies of Dominance

The term "peck order" came into use to describe **social hierarchies** because of the earliest studies that were performed using chickens. There is one individual who usually dominates in any flock of hens and who can peck at any other hen in the group without being pecked in return. A hen who could peck at all of the others except for number one was next in the hierarchy; then a hen who could peck at all except for number one and number two; and so forth. Such peck orders become established by a series of **dominance encounters** through which it is established which hen can beat which of the others in the flock. Such groups are not mere aggregations since it is obvious that each animal in the group can be recognized individually by each of the other members. The hierarchy becomes established gradually but, once it is determined, there is little subsequent fighting since each individual essentially learns its place in the order. A stranger must establish its social place by fighting each member of the dominance-ordered flock.

The social hierarchy in primates and some other mammals includes one **dominant** or **alpha** individual who generally, but not always, is the largest male in the prime of his powers. This position generally is achieved by his overall aggressiveness as well as his physical condition in many species. Once established, he commands the immediate space around him and maintains his dominance by ritualistic threat displays rather than by actual fighting. Other members of the group keep out of his way, do not stare at him, and do not offer opposition to actions he may take. Such strict dominance is more typical of species like the baboon than of other primates that lead a more relaxed existence and correspondingly show relaxed dominance relationships. There is a rank order among the females, too, but the expressions usually are less vigorous than those of the male in the more aggressive species of primates. Some species, such as the chacma baboon, have a clique of dominant males rather than a single alpha individual. This group of dominants acts in concert when defending the group and at the times when they maintain their social position and accept the privileges of rank. In the social groups in which dominance is emphasized and continuously maintained and asserted, the alpha male or males have prior access to all the amenities of their life, including food, females, desirable spaces, and other items they may want.

The intensity of dominance behavior varies among the primate species and generally corresponds to the frequency and extent of the dangers to which the social group is subject. Some monkey species, such as baboons and macaques, which live in surroundings of constant danger from predators, display a considerable amount of aggressiveness in asserting their social rank within the group. Other species such as tree-dwelling monkeys or the gorilla and chimpanzee generally show a relatively relaxed dominance and mode of life. There is little violence, although the dominant male(s) may lead the group in foraging for food and alert the group to danger, at which point everyone scampers for the trees on his own. The studies of free-living gorillas, chimpanzees, and some of the monkeys have clearly shown that there is no consistent, prior access to mates for the dominant males among gorillas and chimpanzees, as there is among the fiercer and more quarrelsome monkeys such as the baboons.

The principal adaptive advantage of dominance hierarchies lies in the fact that such social behavior *stabilizes* the group and provides for *cohesiveness* among the individual members. Without such hierarchies there would be a disruption of order and discipline, which would have especially disastrous consequences for baboons and other species that face constant danger in their daily existence. The advantage of group protection thus underwrites the cohesiveness and maintenance of the society, and not sex as it was once believed. Recent studies of free-living monkeys and apes in the wild have provided us with a more accurate picture and understanding of primate behavior than had been presumed from earlier studies in which caged animals were placed in bizarre and unnatural situations or environments. The anthropomorphic interpretations of some kinds of experiments have been reexamined in the light of more objective and long range observations made in wild populations, such as those of George Schaller for the mountain gorilla and Jane Van Lawick-Goodall for the chimpanzee.

Territoriality

Another adaptive feature of group sociality is the behavioral pattern that leads to the spacing out of **domains** or geographical **territories**. In many species there are encounters between males; or marking of particular spaces by chemical, auditory, or visual signals; or some kind of threat display that spaces out the breeding pairs or carves out an area in which the social group can secure its food and maintain its particular membership. The size of the territory may be as small as the diameter of a circle from which a nesting gull can peck at another without stirring from her nest in the colony; or the three to six square miles occupied by a troop of baboons. Herbivores generally occupy smaller territories that sustain their grazing or browsing habit, whereas large carnivores may range for many miles to secure the necessary food supplies for existence.

Breeding animals frequently establish a territory and aggressively prevent competitor males from invading the space. The males of many songbird species advertise their ownership and warn away other males as well as attract females to the territory. Such territorial behavior provides one means for population control since the spacing leads to greater probability of an adequate food supply for the prospective family as well as limiting the number of breeding individuals when territories are not available. Once territories have been established, there is little competition or fighting so that each breeding pair or social group can devote maximum efforts to the daily business of living. Both territoriality and dominance hierarchies provide means for organizing societies with the least amount of disruption by aggressive actions between individuals. Indeed, the dominance hierarchy can be viewed as a kind of territoriality, in that individual members of the group occupy particular spaces and positions within the society. These occupations are sometimes brief and rather temporary, but they also serve to space out the individuals and minimize disruption of the social order.

Social Organization

Most organisms live together in the same environmental regions because of different needs that are met by different materials that are available in the same area, or occasionally because there is more than enough food for animals with the same nutritional patterns. The organization of individuals inhabiting a space may be societal, aggregational, or random. We will concentrate on the organization of individuals in societies and illustrate some of the variety with selected examples.

Insect Societies

There is a variable spectrum of sociality among the insects, which ranges from solitary existence to the highly stratified colony systems of the social bees, ants, and termites. Among solitary insects, there is in turn a gradation of behavior in parent-progeny relations so that mosquitoes deposit their eggs in any wet or damp place and take off without providing further for the progeny; butterflies lay their eggs on food supplies that will be suitable for the young larvae after hatching; and the solitary digger wasp digs a hole, seeks a specific prey that it paralyzes but does not kill, stocks the hole with such provisions, lays an egg on top of it, seals the nest carefully, and flies away. Some kinds of insects store food for the next generation and remain to guard the eggs, leaving only after hatching has occurred.

The true social insects form matriarchal colonies in which all the members are hatched from unfertilized or fertilized eggs of the same queen or from a very few such fertile females. The individuals are differentiated according to castes, which usually include sterile females that function as workers, and the male and female reproductive castes. Ants and termites also include a soldier class whose individuals possess tremendous jaws and who thus function effectively in protecting the colony. Some ants feed special members of the colony to a point at which these individuals become living honey casks that hang from the roof of underground chambers. Whatever the mode of provisioning may be, the true social insects store foods progressively, and then the young cooperate in caring for the next generation. The great majority of the society consists of individuals that do not reproduce; the business of producing the new generation is restricted to the relatively few males and to the female that becomes greatly distended as she continues her job as an egg-laying machine.

Although insect societies are remarkably complex and coordinated, the structure and organization of the social group is extremely inflexible. The visual and chemical signals that precipitate one activity or another by some or all the members of the society all represent fixed-action patterns of behavior. The rigidity of these behaviors are adaptive under the particular conditions in which the insect colony lives, but they foreclose the possibilities for rapid accommodation to changing environments. The social organization provides for short-term species success and limits the long-term success probabilities that are more likely to occur in species that can modify their behavior in response to the greatest variety of changing conditions. Variability of organization characterizes vertebrate societies, and isolated groups of the same species usually display very similar patterns. The diversity of social organization in vertebrates is greater at the *interspecies* level, except for humans. The human species is unique in possessing an astonishing variety of social behavior patterns, which reflects the increasing *flexibility* of activities made possible by reasoning behavior and a high level of intelligence as compared with other primates and mammals in general. Contrary to some current vogues of thinking expounded by a number of prominent ethologists (students of behavior) and social anthropologists, many human behaviors are learned and reasoned rather than being "programmed in the genes." Cultural influences have come to predominate many aspects of human behavior, which certainly may lead to rigidity of social pattern in some groups but obviously does not lead to identical social patterns in all human societies.

The comparisons often made between insect and human societies tend to emphasize the superficial resemblances and to ignore or dismiss the relative inputs of genetics and culture that sharply distinguish behavioral modes of the two kinds of animals.

Types of Societies in Mammals

Unlike the insects that have a much greater reproductive capacity, the social groups of birds and mammals tend to be smaller and to consist of families or of reproductive groups. The male usually is included in the family in birds but more often is excluded in mammals where the mother and offspring constitute the cohesive unit. There is considerable variety of social arrangement among mammals, and we will briefly examine some representative types that have been well studied.

Along with most mammal species, the red deer has a matriarchal organization that consists of mothers and their young. After three years young males become sexually mature enough to participate in the rutting season and they leave the herd. Young females remain associated with the herd, and as many as three generations of females may remain together in such groups. Mature males live apart from the females in their own loosely organized herds; the female groups are highly organized along dominance-hierarchical lines. The older females guide and protect the herd, and one of these usually exercises a clear position of leadership. At the rut season in the fall the male herds break up and individual males scatter over a wide range, invade female territories, and establish harems. During the breeding season the male maintains a constant and exhausting set of sexual and defensive activities, servicing the females in his harem as they come into heat and aggressively keeping other males from his territory. The exhaustion takes its toll and the stag will depart for the uplands as he is forced to make way for another more energetic male. Foals stay close to their mother from the time they are born in the spring until the first winter at which time they have achieved a considerable level of growth. In this kind of society the males serve only a reproductive function, and the females are responsible for the cohesiveness of their group and for educating and caring for the young. According to some observations, female young are much more "attentive" to the activities of the older females, whereas young males still with the maternal herd do not seem to take as much notice of these goings-on. The introduction to the male pattern of behavior that omits parental, educatory, and defense activities begins when the animal is very young.

The wolf often has been compared with humans since the socialization and family patterns appear similar in many respects. These carnivores hunt large prey in packs, which requires cooperative efforts for success, and they utilize stratagems during hunting that further indicate reasoning and cooperativeness among the members of these groups. The usual pack is small and comprised of members of a family unit, although several packs may combine temporarily to hunt. The males contribute substantially to family life and care for the females who are pregnant or lactating newborn young, as well as for the young themselves. Females participate in hunting except when pregnant or caring for newborns, at which times they are fed by the males that have returned from a hunt. There is an alpha male who has prior access to females; however, a breeding pair may remain together for life or for many years. The mate of the alpha male exercises an alpha status in relation to the other members of the pack, except for her mate. The cooperativeness of pack-hunting species has a clear adaptive value, and it appears to be reinforced by continual social interactions and interdependences among the adults and the young of the social unit. The social organization of animals like the wolf is unusual among mammals in that nonbreeding males are full and participating members of the group. There is very little fighting or other aggressive behavior within the group, and the social bonds formed during play in the preadult years serve to reinforce the family relationships and stabilize the social unit. The young learn from all the adults of the wolf social group in which they develop, and parental-progeny interactions are maintained throughout the life of the members of the family.

Primate societies are extremely varied, ranging from the more rigid social structure of some baboon and macaque species to the more relaxed and open societies of other monkeys, the gorilla, and the chimpanzee. Since apes and monkeys primarily forage for their food, the strictness of the dominance hierarchy and the socialization among members of the group probably bears a fairly direct relationship to the availability of the food supply and the prevalence of predators. Young monkeys and apes are most closely associated with their mothers, for as long as three or four years in the great ape species, and they interact with adult males to a variable extent in the different species. Although there is some group participation in educating the young, the principal interaction is between the mother and her offspring. In species such as most of the baboons, the dominant males have first access to females in estrus and actively prevent other males of the group from mating with receptive females. The dominant male mountain gorilla, on the other hand, has been observed in one study to ignore a male stranger to a group while the stranger approached a female in estrus and then proceeded to copulate with her. Regardless of the specific expressions of dominance and other aspects of social behavior, monkeys and apes demonstrate the existence of a relatively high level of intelligence in the management of social relationships within the group and between groups.

Mating Behavior

The perpetuation of the sexually reproducing species requires that male and female find each other and come together cooperatively, that the eggs and sperm are produced at the same time or in some appropriate conjunction, and that the new generation of individuals are protected at least until birth or hatching and for some time after birth in many species. These may seem like relatively simple requirements when thus stated, but these accomplishments have been made possible by varied and elaborate adaptive mechanisms and devices during the course of evolution.

Meeting of the Sexes

Among the vertebrates a variety of phenomena exist whereby males and females of the same species arrive at an appropriate location where mating can take place. The remarkable migration and directed navigation of the Pacific salmon is well known and may serve to illustrate one situation. The adult fish remain in the sea until near the time for spawning, at which time the five different species of Pacific salmon migrate from the ocean to the freshwater streams where they were hatched: to spawn and die. The accuracy of their return includes finding the precise twists and turns of the tributary streams until the spawning ground is reached. Studies have demonstrated that the young salmon respond to the stimuli of particular odors of the different streams where they were hatched, thus finding their way back almost unerringly. Atlantic salmon return each year to their spawning grounds until accident or old age overtakes them, unlike the Pacific species that spawn only once in their lifetime.

Most vertebrate and invertebrate animals respond to signals sent by one of the partners, which may be chemical, visual, or auditory in the commonest systems. Moths are well known to produce **pheromones** which are chemical secretions involved in communication within the species, such chemicals being produced in special abdominal glands. These highly volatile secretions attract males, requiring as little as one molecule reaching each olfactory cell every two seconds to turn the males toward the female source of the pheromone. Pheromones also are secreted by vertebrates from variously deployed glands near the eyes, horns, tail, or around th anus; and they may also be deposited in the wastes eliminated from the body. Such chemicals serve as territorial markers and orientation signals as well as sex attractants. Anyone who has experienced the nuisance of tomcats spraying (urinating) on the car or furniture has witnessed a marking of territory.

Sounds are important signals among insects, amphibians such as frogs and toads, and birds. The songs (stridulations) of some crickets and grasshoppers are signals by the male and are recognized discriminately by the female of each species. Songs of male frogs and toads similarly bring females of the proper species to the breeding grounds in the spring. The distinctness of such songs virtually ensures mating between males and females of the same species, thus preventing excessive reproductive waste and inefficiency of hybridization between genetically incompatible individuals.

Fish and birds respond especially to visual signals, the zigzag dance of the male stickleback and the posturings of the male and female lead to the consummation of mating activities. The numerous descriptions of courtship display among birds, especially those that inhabit open spaces, attest to the evolutionary significance of visual signaling to prospective partners. The posturings of male peacocks can be observed in many zoos, along with the indications of acceptance or disdain by the solicited females. It is usually the female of the species which selects a mate in these animal groups.

Coordination of Reproductive States

Once males and females have met for purposes of reproduction, success of the species requires that they first have acquired an appropriate state for mating. Thus we expect that eggs and sperm have been produced or can be produced in a proper conjunction of times; that is, there usually is a synchrony of time and of place for gamete production. This usually is accomplished by stimuli from the external environment or from members of one sex upon the other.

In many vertebrates there is an effect of increasing daylength on the pituitary, which in turn is caused to release gonadal hormones that affect behavior. Hormonal changes lead to other modifications that finally result in reproductive behavior. In mammals, the female in estrus, as a result of estrogen secretion during the midpoint of the cycle, assumes a mode of behavior that signals her reproductive state and readiness to the male. In species that breed during specific seasons, whether once or many times in the season, the males manufacture sperm throughout the breeding period and can inseminate the female to achieve successful fertilization. Species in which ovulation is induced by copulation with the male are ones in which fertilization has a particularly high chance for success. But there are various other mechanisms by which mating success is achieved and which also serve to ensure species continuity if other life conditions are appropriate.

The human species is unique among the primates in lacking a precise coordination of ovulation and reproductive behavior. The lack of a defined estrus in the human female, together with biological capacity for mating behavior at almost any time by either sex, represents an exception among animals.

Care of the Young

In many species there are family or communal units within the social group that are comprised of male and female adults, juveniles, and young. This sociality is a consequence in part of sexual behaviors of the adults, the general defense and protection afforded by the group, and relations between parent(s) and offspring. The general basis for family life revolves around the shelter, feeding, and protection of the young; and these individuals provide for the continuity and success of the species. For such a family relationship to remain viable there must be appropriate modifications of behavior as circumstances change with time. There must be an appropriate suppression of otherwise "natural" responses if the family's relationship is to succeed. The young must not view the adults as predators nor must the adults consider the young to be delicacies for consumption. Just as the adult is stimulated to feed the young by one signal or another, such as a particular pattern of spots that signifies the gaping mouth of a young bird in the nest, the young also are stimulated to take food or beg for food by shapes and colors or actions of the parents.

For any group to persist and function, whether comprised of members of the same species or of different species, there must be a continual flow of information and assimilation of this information. Chemical or visual signals may serve to induce aggregation of individuals, but cohesiveness and coherent functioning require a steady input of information in successful social groupings.

The formation of a group provides greater insurance against predators and other dangers in the surroundings of the animals. In mixed-species groups such as those on the African plains, each species performs some important function that serves a common purpose for the entire assemblage. If antelopes and baboons occupy a common space, the baboons can sight danger because of their acute vision while the antelopes can smell danger otherwise not visible to the monkey species. Each serves the other, and together they comprise a more successful gathering for each species than either could accomplish alone. Within a species there are numerous examples of protective displacement of the adult individuals such that the young are within the center of the group and the males are distributed around the perimeter in a defense posture. Females in estrus and females with infant offspring are afforded greater protection in some cases, as with baboons, than is true for the other adult females although all the females are within the common defense perimeter (Fig.

9.1). Such activities provide a system of defense that protects the least dispensible members of the group—the females and the young individuals of the next generation. Males are more expendable since a few males can serve as sex partners whereas only the female can bear the young, and each female contributes an essential component for continuity... the offspring of the species.

Development of Behavior

The extensive studies of Rhesus monkeys and other macaques conducted at the Primate Laboratory of the University of Wisconsin by Harry F. and Margaret K. Harlow and their associates have provided us with numerous and important insights into the development of some patterns of primate

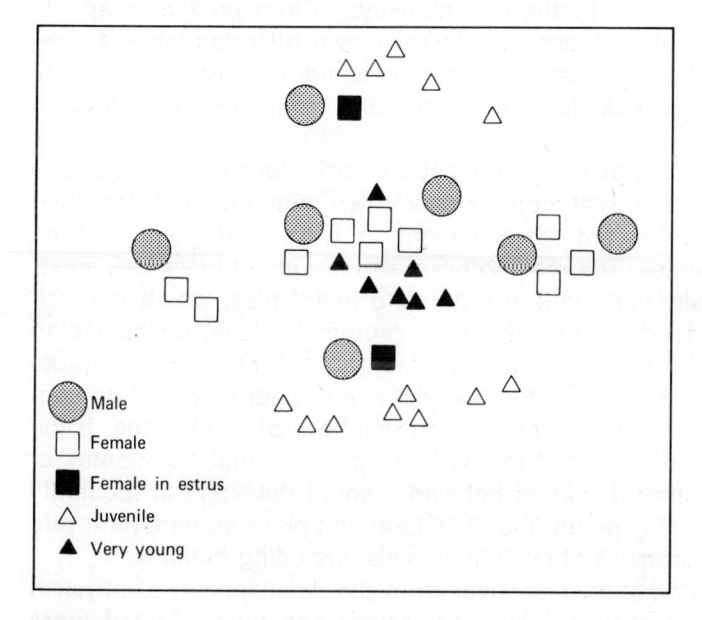

figure 9.1
Spatial pattern of a troop of baboons on the move. A group of dominant males, adult females, and young occupy the center. Other members of the troop, including juveniles, are distributed concentrically. Two of the females are in estrus and have consort. (Modified from *Primate Behavior Field Studies of Monkeys and Apes*, edited by I. DeVore, 1965, Fig. 3-10, p. 70. Copyright © 1965 by Holt, Rinehart, Winston, Inc. Reprinted with permission of Holt, Rinehart, Winston, Inc.)

behavior and influences over the development of these patterns from infancy to adulthood. Underlying the development and learning of social and reproductive behaviors is the sequencing and interaction of five major affectional systems, which the Harlows have termed maternal love, love of the infant for the mother, peer love, heterosexual love, and paternal love.

Interactions Between Mother and Infant

Babies that had been separated from their mothers a few hours after birth were raised in individual cages under human care and provided with different living situations. Surrogate mothers covered with terry cloth were preferred to wire mothers even if the latter provided the only source of milk, showing that reassuring contact was an important developmental variable. Continued lack of a mother or surrogate led to the development of totally abnormal monkeys incapable of interacting with other monkeys of any age or sex. These motherless monkeys in turn proved incapable of providing maternal care and love to their own offspring when forced matings were arranged so that such animals could bear their own young. Motherless monkeys practiced cruelties and even killed their own infants.

Although surrogate mothers could instill a basic sense of security and trust in young monkeys, real mothers performed numerous activities that helped the total behavioral development in ways that could not be achieved by surrogates. The real mother trains her infant in various ways, but she is especially important in regulating infant play, which is a primary activity that leads to effective peer (age-mate) love. By managing infant play in an effective rather than a disorganized manner the expression of maternal love leads to the development of peer love and its consequent social interactions among young monkeys of similar age. If the infant remained totally isolated for no more than the first six months of its life, normal or almost normal behaviors could develop subsequently within a social or family group (Fig. 9.2). Learning plays an important role in behavioral development of many mammals, including humans.

The influence of emotional deprivation in the development of physical and psychological problems in human infants has been studied more carefully in recent years. Information derived from studies of neglected children has shown that a hostile or unresponsive environment may cause severe disturbances in development even if the infant is well nourished. A high fatality rate occurs especially among infants between the seventh and twelfth months of life if they are emotionally deprived, and severe physical retardation characterizes many who survive beyond their first year. The effects of emotional deprivation have been associated with

Sexual and Sex-Related Social Behavior 199

Experimental condition	Present age	Behavior* None	Low	Almost normal	Normal
Raised in isolation					
Cage—raised for 2 years	4 years	●○◎			
Cage—raised for 6 months	14 months	○◎	●		
Cage—raised for 80 days	10 months			●○◎	
Raised with mother					
Normal mother; no play with peers	1 year	○	●		○
Motherless mother; play in playpen	14 months			○	● ◎
Normal mother; play in playpen	2 years				●○◎
Raised with peers					
Four raised in one cage; play in playroom	1 year				●○◎
Surrogate—raised; play in playpen	2 years				●○◎
Surrogate—raised; play in playroom	21 months				●○◎

* ● = play,   ○ = defense,   ◎ = sex

figure 9.2
A summary of some of the results from experiments reported by Harlow and Harlow showing the importance of learning and of interactions between individuals for the effective development of social relations and species behavior in the Rhesus macaque monkey.

modifications in the pituitary secretions of hormones, especially of the growth (somatotrophic) hormone. The relationship between pituitary abnormalities and influences from the hypothalamus and cerebral cortex are now under active study. Infant-parent interactions thus influence behavioral development in various delicate and critical ways, only one of which involves the processes of learning.

From these and other studies of human infants and from experimental observations of young Rhesus monkeys it is quite clear that the infant obtains needed love, a sense of security, and education from its mother. A young monkey or human child will undertake explorations of its immediate surroundings in the presence of a mother, whether real or surrogate. The infant returns repeatedly to the mother for assurance and then continues to explore in wider and wider circles as its sense of security is established. Youngsters who are alone will retreat to a corner, and show little activity except for terror and bewilderment.

Interactions Among Peers

According to the Harlows, peer love may be the most important factor in the whole life of the individual Rhesus monkey. Age-mate love develops first through curiosity and exploration and is enhanced by various forms of play later on. Interactions between peers enhance a whole array of developing behaviors including the formation of affection for age-mates and other associates, basic social role development, inhibition of aggressive actions, and the maturation of basic sexuality. Heterosexual love probably evolves from peer love systems, but it differs in form and function in different animal groups. Many mammals experience a sex life that is determined largely by neuroendocrine systems, whereas primate heterosexuality is subject to more complex variables and controls or flexibilities. Learning forms a more substantial part of the development of heterosexual behavior in primates. Monkeys that have been raised under one or more regimes of deprivation may never achieve successive heterosexuality. Monkeys that have grown up under normal conditions of social life still must learn to participate successfully in sexual activities, and this is accomplished through practice, experience, and learning within the social and family groups themselves.

Paternal Love

The role of the adult male in the development of infant behaviors has not been explored as thoroughly, but there is information from studies of free-living primates and other mammals, as well as experimental observations from Primate Laboratories such as the one at the University of Wisconsin. Since, in an earlier section of this chapter, we briefly discussed the varying role of the male in several species' social groups, we know that different patterns do occur.

In the experimental situations described by the Harlows, macaque adult males belonging to an interconnected cage community of adult pairs and offspring showed cohesive behavior in guarding the group against the human experimenters (predator surrogates), and they did not allow abuse or abandonment of infants belonging to any of the adult members of the small group. The fathers ignored aggressions from infants and juveniles that they would not tolerate from adults of either sex or even from adolescent monkeys. They engaged in reciprocal play activities with infants to a much greater extent than the Rhesus mothers. The development of paternal love probably begins during preadolescence when juvenile and even older male infants will cradle, carry, and protect young infants in their path, presumably under the watchful and threatening eyes of the adults. Preadolescent females appear to be more motivated toward new babies, although preadolescent Rhesus males tend to exhibit less interest until the babies are old enough to engage in play activities. Once the infant is old enough to play, the interactions between infants and young nonadult male monkeys may contribute to the development of a protective behavior pattern. The forms of play in Rhesus monkeys appear to be sex-typed rather than imitative, that is, to be based in large measure on genetic and/or hormonal factors. This interpretation received support from studies of rehabilitation in socially deprived monkeys raised in isolation for the first six months of their lives. Six-month-old males were removed from isolation and associated with three-month-old female "therapist" monkeys. Normal Rhesus male monkeys develop a more contact-oriented and rougher play repertory than females, and they have a well developed play behavior by the age of six months. Isolated six-month-old Rhesus males developed a clearly masculine form of play, once it had emerged following "therapy," even though there had been no models on which this particular development could have been based. In addition, the abnormal behaviors completely disappeared after six months of "therapy" so that the previous isolates assumed essentially normal activities by the time they were one year of age. The implications for humans of the numerous observations regarding the development of behaviors, psychopathology, psychotherapy, and rehabilitation in Rhesus monkeys remain to be explored.

From these studies and many others it is clear that young primates and other mammals acquire behaviors on the basis of experience and learning, as well as imitation, superimposed on the background of their genetic endowment. Sensory and psychological deprivations lead to abnormal behaviors when young animals are subjected to such exposures during their early formative years. Some of these abnormalities may not be correctable even when normal surroundings are provided at a later time during development. The monkeys learn to become participating members of a social group and can reproduce only if they have learned the proper procedures and have perfected these methods by practice and experience. In free-living monkeys and apes there is ample opportunity to learn, to imitate, to practice, and to gain the experiences by which the youngster can take its place in the adult society; thus the youngster contributes to its daily existence and to the perpetuation of the species. The degree of similarity between behavioral development in humans and subhuman primates forms one basis for controversies in interpretations and notions put forth by different camps of biologists, psychologists, and anthropologists. We will explore some of the controversies in the next group of chapters.

Suggested Readings

Andrew, R. J., The origins of facial expressions. *Scientific American*, 213 (October 1965), 88.

Berkowitz, L., Simple views of aggression. *Amer. Scientist*, 57 (1969), 372.

Calhoun, J. B., Population density and social pathology. *Scientific American*, 206 (February 1962), 139.

Christian, J. J., and D. E. Davis, Endocrines, behavior, and population. *Science*, 146 (1964), 1550.

Davidson, E. H., Hormones and genes. *Scientific American*, 212 (June 1965), 36.

Dolhinov, P., At play in the fields. *Natural History*, 80 (December 1971), 66.

Guhl, A. M., The social order of chickens. *Scientific American*, 194 (February 1956), 42.

Harlow, H. F., Love in infant monkeys. *Scientific American*, 200 (June 1959), 68.

Harlow, H. F., and M. K. Harlow, Social deprivation in monkeys. *Scientific American*, 207 (November 1962), 136.

Harlow, H. F., M. K. Harlow, and S. J. Suomi, From thought to therapy: Lessons from a primate laboratory. *Amer. Scientist*, 59 (1971), 538.

Hasler, A. D., and J. A. Larson, The homing salmon. *Scientific American*, 193 (August 1955), 72.

Johnson, C. E. (ed.), *Contemporary Readings in Behavior*. New York: McGraw-Hill, 1970.

Levine, S., Sex differences in the brain. *Scientific American*, 214 (April 1956), 84.

Levine, S., Stress and behavior. *Scientific American*, 224 (January 1971), 26.

Morris, D. (ed.), *Primate Ethology*. Chicago: Aldine Publ. Co., 1967 (reprinted by Doubleday in paperback, Anchor A692, 1969).

Singh, S. D., Urban monkeys. *Scientific American*, 221 (July 1969), 108.

Suomi, S. J., and H. F. Harlow, Monkeys at play. *Natural History*, 80 (December 1971), 72.

Tajfel, H., Experiments in intergroup discrimination. *Scientific American*, 223 (November 1970), 96.

Washburn, S. L., and I. DeVore, The social life of baboons. *Scientific American*, 204 (June 1961), 62.

Watts, C. R., and A. W. Stokes, The social order of turkeys. *Scientific American*, 224 (June 1971), 112.

Wilson, E. O., Pheromones. *Scientific American*, 208 (May 1963), 100.

Weisz, P. (ed.), *The Contemporary Scene*. New York: McGraw Hill, 1970, pp. 131–277.

chapter ten

Human Sexual Behavior

The meaning of "sexual" behavior as opposed to other manifestations of adjustive responses to external stimuli essentially is defined in terms of reproductive activities for all species except our own. For humans there is usually a connotation of the romantic and erotic, but less often the implication of reproduction. These different outlooks are reflected in the ways in which we come to analyze sexual behavior in humans as compared with other animal species. A study of sexual behavior in the lion consists of information concerning frequency of copulations, the receptive or aroused states of the copulants, and the success of the copulatory acts in procreation. In addition there are data to be collected concerning dominance, numbers of females in the average harem, neuroendocrine correlates of sexual behavior and the participants, and so forth. We actually know a great deal more about sexual behavior in other vertebrates than we do about the human species, precisely because specific correlates can be measured. An entirely different measure has been used for analyzing human sexual behavior, not only because the data can be obtained directly and simply but also because the data can be quantified or measured on an "objective" basis. One of the more frequently used measures of human sexual behavior and expression is *orgasm*. The detailed studies of the Kinsey group, beginning in 1948 with *Sexual Behavior in the Human Male*, provided the basis for recognizing the principal means by which men and women achieve orgasm. The Kinsey studies showed that there were six major methods by which orgasm is attained: masturbation, coitus, heterosexual petting, sex dreams, homosexual relations, and animal contacts. Various other activities may occur independently or in conjunction with one of these major methods, although they tend to be rare. Examples of these would include sadism, fetishism, and other experiences of this general type.

The Kinsey reports were the first large-scale studies that involved thousands of people of different but selected groups. Although many individuals contributed to the data that were collected, these studies were not sufficiently representative of the population in the United States and certainly not for the whole world of human beings. But they did provide a reasonable sampling of the groups that were the subject of each of the separate reports. In addition to criticisms concerning sampling and therefore of extrapolating the information to the species as a whole, there were other criticisms of these reports. Cross-cultural comparisons were not adequate, which further limited the generalizations that could be derived from the studies; and the technique of personal interviews with voluntary subjects raised questions concerning the objectivity and accuracy of the collected information. Notwithstanding these sound lines of evaluation, the Kinsey reports represented an innovative milestone in the study of human sexual behavior. A variety of other reports followed, including the classical cross-cultural studies published by C. S. Ford and F. A. Beach in 1951, and the more clinically oriented publications in 1966 and 1970 by W. H. Masters and V. E. Johnson that provided previously unavailable information concerning the physiological concomitants of sexual intercourse. These books and others have provided the main sources of information contained in this chapter.

Measurements Versus Assessments

Even though a great deal of the available information has been tabulated, averaged, and clothed in respectable mathematical terms, there really is no basis for knowing the manners and distributions of the patterns of sexual behavior in most of the world population. However, the studies of relatively small groups of people do permit us to see immediately that varied sexual behaviors are characteristic of humans as a species. Furthermore, we can tell that variability is present in different communities and among different individuals of these communities. In accord with every other known aspect of human behaviors, our sexual activities also reflect the uniqueness of human beings as the most variable species in the animal world. The variability and complexity of sexual and other behaviors reflect the pervasive influences of social and cultural factors on the expressions of the genetic potential. A species that can successfully invade and adapt to a startling variety of living zones and styles is one that must exhibit behavioral plasticity based on intelligent exploitation of its genetic potential in many environments and situations. But because of cultural influences on our biological activities, our behaviors are evaluated according to subjective judgments of our own creation.

In seeking to establish a norm and thus to derive one criterion of judgment, we not only ask how common is the behavior pattern, but we also create other criteria based on other questions that we pose. We impose the criterion of healthy behavior in seeking a medical judgment; a moral behavior to satisfy an ethical judgment; and a legitimate behavior to satisfy a legal judgment. Each of these criteria may be applied differently in different societies or even in the same society at different times. Each of these criteria may conflict with one or more of the others, thus producing contradiction and confusion. None of these criteria represents an absolute value, and none of these criteria receives support from any convincing body of evidence or general consensus. A child molester may be regarded with abhorrence as violating a moral code, or sympathetically as a victim of unsound health, or dispassionately as a criminal according to a legal code. We all would agree however that such a person was not typical because the particular behavior is relatively uncommon as a form of sexual expression in human societies. But what does it mean to be "typical" in one's sexual behavior? Our societies are pluralistic, and our species is typically variable. To seek the norms for behavior or to determine the relative proportions of the different expressions of sexual behavior we must employ statistics, knowing very well that a norm is merely an average, and that a range of variability usually occurs on both sides of the arithmetical mean value (Fig. 10.1).

Statistical Information

The major source of our statistical information comes from the 5300 males and 5490 females whose sexual histories were established from personal interviews by A. C. Kinsey and his staff. These data were collected one generation ago and may not be as applicable in the United States today as when they first were published. Also, as mentioned earlier, the sampling is not only small but also represents a highly selected group of white Americans. There were subjects of varying ages, occupations, religions, educational levels, and so forth, but the whole span of white Americans certainly was not included. Since only one racial group was included in the published reports, a significant component of the population was not utilized in the constructs and general-

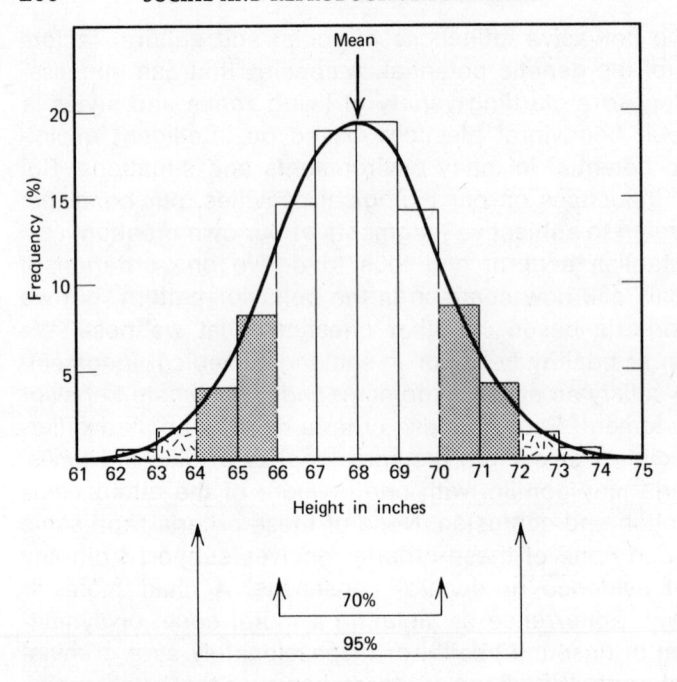

figure 10.1
Height distribution among a sample of American men. In this multigenic trait as in others, such as weight, intelligence, etc., there is a range of variation that exhibits a symmetrical "bell-shaped" distribution curve. It is meaningful to discuss a *mean* (arithmetical average) height, since about 70 percent of the sample deviates no more than two inches from the average and 95 percent of the men deviate no more than four inches from the mean height. Those 5 percent that differ by more than four inches from the average are unusual, but certainly not "abnormal," in that they are found least frequently.

izations that emerged. Despite these obvious shortcomings, the Kinsey reports are a major source of information amenable to statistical analysis of human sexual behavior.

The particular measure used by Kinsey to determine a person's sexual activities was the number of orgasms attained through one or more of the six major methods that were listed earlier. This expression provides an objective datum that can be tabulated and quantified, whereas other manifestations of sexual behavior are more difficult to analyze objectively. The frequency of orgasms was calculated on a weekly basis, since numbers *per se* have little comparative dimension and we require a temporal basis for standardization in data interpretations. Another component of the statistical treatment involved minimizing the variability of individual

Frequency of Orgasm in Men and Women

Among males between the ages of adolescence and 85, the range in frequency of orgasms extended from 0 to 30 per week, based on five-year averages. The weekly average per male equalled three orgasms, but this number is less meaningful when the total distribution of frequencies is examined (Fig. 10.2). One-third of all the subjects experienced one or two orgasms per week; a total of three-fourths attained from one to six orgasms per week; and the remaining one-fourth were outside this range of variation. The arithmetical average is higher because the higher frequencies of relatively few individuals unduly influence the calculations. The distribution pattern indicates that the most common classes fall below three orgasms per week, but that an extensive variation exists.

The most important factor that contributed to orgasmal frequency in men was their age (Fig. 10.3). The most active group included individuals ranging from postpubescent boys less than 15 years old to men up to 30 years of age. There was a steady decline after this, but a somewhat unexpected observation was that no essential difference in frequency of orgasm distinguished any age group under 30, since approximately the same values were characteristic for the youngest postpubescent boys as for any adult age group up to 30 years old. Contrary to popular notions,

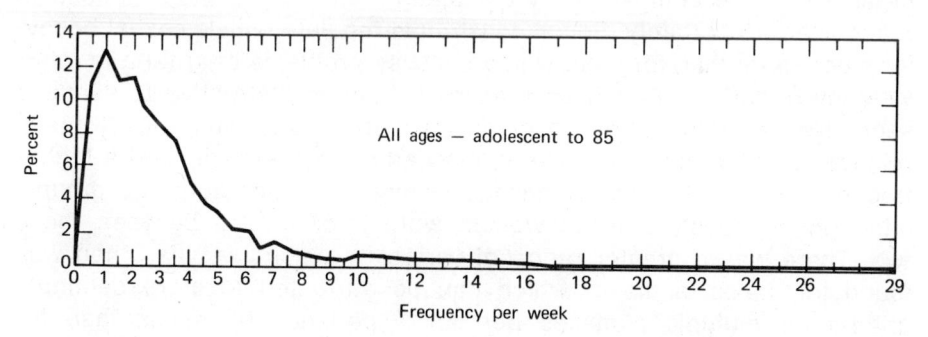

figure 10.2
The frequency of total sexual outlet among males ranging in age from adolescence to 85 years. There is substantial individual variation, and the mean frequency of three orgasms per week is relatively meaningless in view of the skewed distribution. The individuals showing a high frequency contribute to the *arithmetical* average more substantially, even though such men were very infrequent in the sample interviewed. (From A. C. Kinsey and colleagues, *Sexual Behavior in the Human Male*, p. 198. Philadelphia: Saunders, 1948. Published with permission of the Institute for Sex Research, Inc.)

figure 10.3
Comparison of the mean frequency of total sexual outlet in males and females, by age groups. (From A. C. Kinsey and colleagues, *Sexual Behavior in the Human Male*, p. 220, 1948; and *Sexual Behavior in the Human Female*, p. 548, 1953. Philadelphia: Saunders. Published with permission of the Institute for Sex Research, Inc.

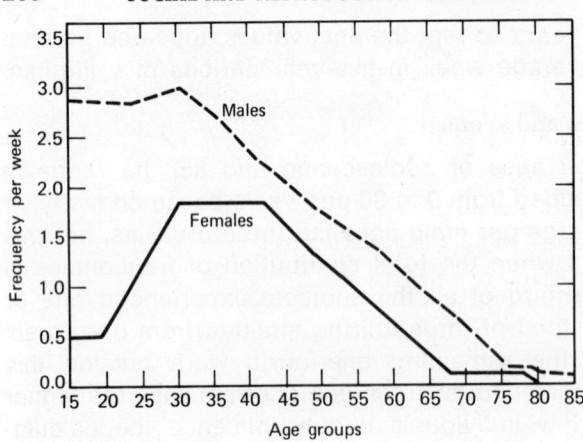

there is no gradual attainment of a "prime" with respect to this *one* level of measurement.

The data collected for females were of a more subjective nature since orgasm is not accompanied by ejaculation and thus is more difficult to recognize with certainty. Social constraints regulate female sexual behavior much more than for men, which imposes problems of criteria in interview investigations. For these reasons it is more instructive to examine separately the information collected for married and unmarried women, relative to age groups. In general, married women experienced a higher frequency of orgasm that unmarried women in comparable age groups, although previously married women were intermediate between these two. There was a greater range of variability among females than was found for males, some of which can be attributed to more definable criteria for multiple climaxes per sex experience by males than for females who were interviewed. The weekly averages were lower for females than for males in all comparable age groups.

The more significant correlations that emerged from these studies were the correlations reflected in the patterns of relationship between age and the frequency of attaining orgasm (Fig. 10.3). Whereas the male group of adolescents between the time of puberty and 15 years of age were as sexually active as males between the ages of 15 and 30, and all exhibited a peak rate during these years, postpubescent females up to the age of 15 were essentially inactive sexually and they rarely achieved orgasm.

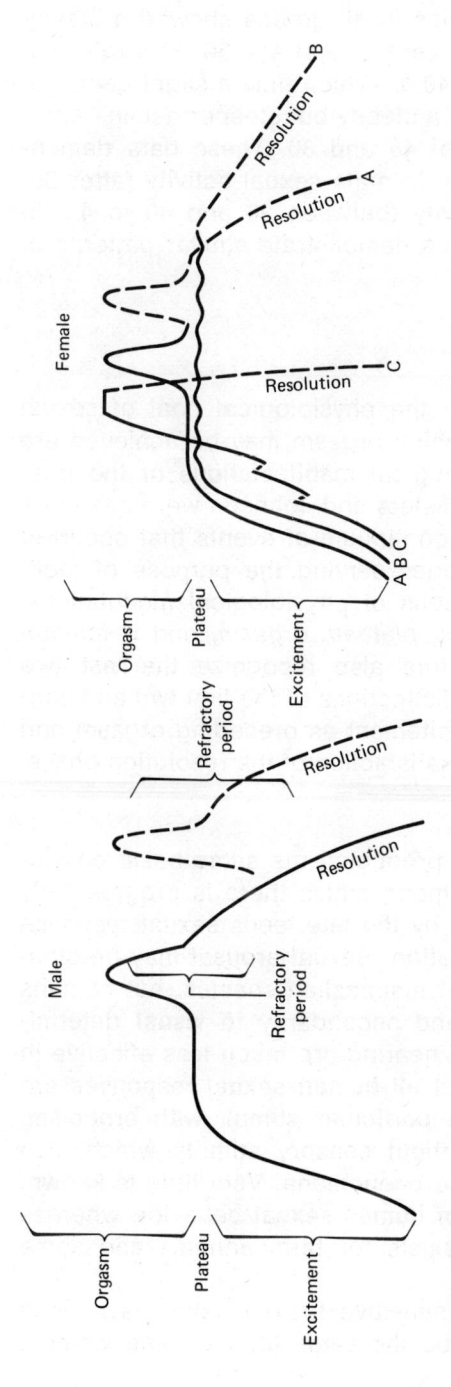

figure 10.4
The sexual response cycle in males and females. (From Masters and Johnson, *Human Sexual Response*. Boston: Little, Brown and Company, 1966, p. 5. Reprinted with permission.)

Between the ages of 15 and 30, females in all groups showed a steady increase in frequency of climaxes to a peak rate at age 30. This rate was maintained at a steady level until age 40, at which time a slight decrease occurred for the 40–45 age group, and a steady but steeper decline characterized women between the ages of 46 and 80. These data demonstrate that the average time of decline in male sexual activity (after 30) coincides with the time of peak activity (between 30 and 40 to 45) in women. After the age of 45, both sexes demonstrate similar patterns of declining frequency of orgasms.

Physiological Concomitants

The attainment of orgasm constitutes the physiological goal of sexual activities in humans. The means by which orgasm may be achieved are independent of the particular physiological manifestations of the phenomenon. The studies published by Masters and Johnson were based on the recognition of four phases in the continuum of events that occurred during sexual activities, these categories serving the purpose of facilitating the observations and descriptions of physiological modulations. Their phases were termed *excitement, plateau, orgasm,* and *resolution* (Fig. 10.4). Although most investigators also recognize the last two phases, some are inclined to blur the distinctions of the first two and consider a general phase of mounting excitement as preceding orgasm and the subsequent relaxation and sexual satisfaction of the resolution phase.

Sexual Stimulation and Response

All healthy individuals are capable of producing the same basic physiological response to varying stimuli. Among males there is progressively more discrimination of stimuli so that by the late teens sexual response is more dependent on physical stimulation. Sexual arousal may be stimlated by any of the senses in different mammalian species, but humans primarily respond to **tactile** stimuli and secondarily to **visual** determinants. The senses of smell, taste, and hearing are much less effective in evoking arousal responses, but almost all human sexual responses are conditioned by learning to associate particular stimuli with eroticism. Emotional stimulation is possible without sensory stimuli, which also implicates the nervous system in these phenomena. Very little is known, however, about the neurophysiology of human sexual behavior, whereas a considerable body of knowledge exists for farm animals and some other species.

Some parts of the body are more sensitive to touch and have been called **erogenous zones**. These may be the same for men and women, such as the mouth, ears, buttocks, anus, and inner surface of the thighs, among other regions; or different in relation to secondary sex characteristics, such as the glans penis in males and the clitoris and inner labia in females. There appears to be no correspondence to particular patterns of nerve distribution in all cases, and neural influences of previous experience and conditioning probably predominate.

In accord with most aspects of human behavior there is variation in the sexual response patterns so that norms or standards have little value in the interpretations. The progression of the response may vary from rapid and relentless, to uneven, to gradual; but there are four sequential phases in the sexual response cycle for both males and females according to the information amassed by Masters and Johnson from analysis of 10,000 orgasms. The type of stimulus, intensity of response, and duration of the different phases are variable; the sequence is not variable.

Orgasm in physiological terms involves a relatively explosive release of accumulated neuromuscular tensions. The onset of orgasm is more distinct than its termination and, in many mammalian species (but not all), the female in estrus may remain responsive to stimulation after

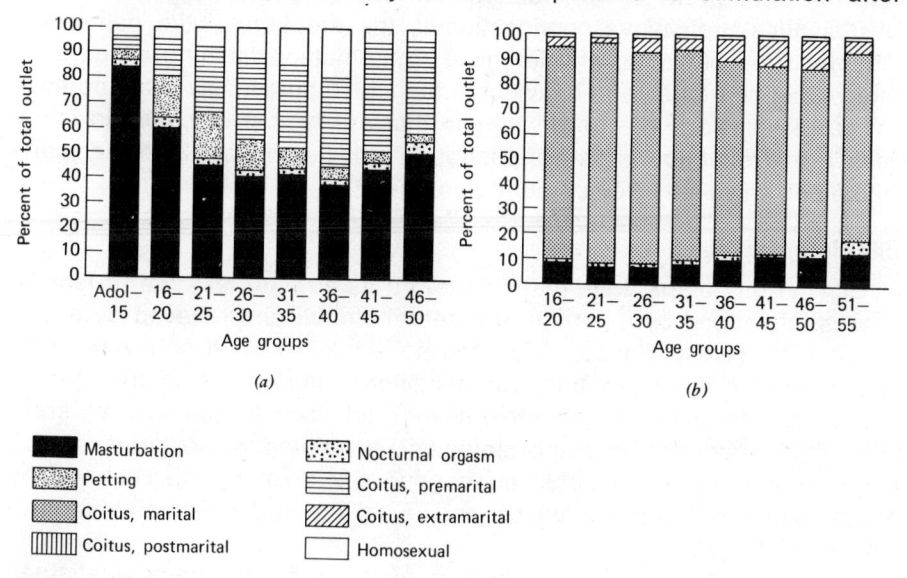

figure 10.5
Sources of orgasm expressed as percentage of total sexual outlet among females, by age groups: (a) single, and (b) married. (From A. C. Kinsey and colleagues, *Sexual Behavior in the Human Female*, p. 562. Philadelphia: Saunders, 1953. Reprinted with permission of the Institute for Sex Research, Inc.)

coitus whereas the male often loses further interest. There are some fundamental differences in response between men and women (Fig. 10.4) according to the Masters and Johnson study among others. Women exhibit greater variability and variety than has been found for male responses and pattern. Largely because of anatomical and physiological differences, women may attain repeated orgasms in rapid succession, perhaps as many as 10 to 12 in an hour, whereas this is impossible for males. The refractory period between successive climaxes is much longer for human males than for females, in general. George Schaller's observations of the lions of Serengeti National Park in north central Tanzania revealed that the male may copulate with the female 60 times in a single day, each event perhaps lasting only half a minute.

The physiological mechanisms of sexual response are **vasocongestion** and **myotonia**. Vasocongestion refers to an increased flux of blood into the tissues subsequent to the engorgement of blood vessels. There usually is a balance between outflow of blood through the arteries and the veins, but during arousal there is more bloodflow via the arteries and less flow out through the veins, thus leading to engorgement. The most obvious manifestation of vasocongestion is the erection of the penis (see Chapter five), but other physiological signs include flushing of the skin, more rapid heartbeat, and heavy breathing. Myotonia involves an increase in muscle tension. The pelvic thrusts that are typical of coitus in mammals are a reflection in part of change in muscle tension, but other parts of the body also are affected during arousal.

Sex Hormones and Sex Drive

In clinical terms **sex drive** may be viewed as an internal state of tension that is influenced by a variety of external stimuli and relieved by a particular kind of experience. This does not explain sexual behavior, but it has merit for the description of behavioral modes. As is true for so many aspects of nonreproductive sexual activities in humans, we know very little about the neurophysiological correlates to sex drive. Much more information is available concerning sex drive in fish such as the three-spined stickleback, which seems to be under complete neuroendocrine control.

It has been documented that human adult sex drive remains unaffected by removal of the gonads, whether testes or ovaries, although these organs are the main sites of secretion of the principal androgens and estrogens. Because it is virtually impossible in our present state of limited knowledge to distinguish genetic from learned and experiential influences in promoting physiological responses, we cannot evaluate the few data that do exist for humans. The responses of women to exogenous testosterone and for men to exogenous estrogens are so variable that no conclusions can be drawn from the available data. Although there is an amelioration of specific secondary sexual characteristics and an enhancement of some traits more usually associated with the opposite sex, no known evidence exists that points to a correspondence between the sex hormone displays in men and women and their particular patterns of nonreproductive sexual activity and behavior. Except for mythologies of popular notion, there is no basis for believing that men behave more aggressively than women or have greater sex drive than women, or vice versa. Neither can we extrapolate to human sexual behaviors the observations showing that testosterone injections lead to more aggressive malelike behavior in hens while exogenous estrogens produce more femalelike behavior in roosters. Chickens and other birds are genetically and developmentally very different from mammalian species in their physiological concomitants of behavior. In other words, we are not chickens.

Varieties of Human Sexual Experience

Although the occurrence of various modes of sexual gratification had been known since antiquity, the extent of the utilization of each mode was not known until the presentation of quantitative data by Kinsey and his colleagues. Before briefly discussing selected aspects of these data, it is necessary to describe the six major modes of achieving sexual arousal leading to orgasm as these were included in studies by Kinsey and others.

Self-stimulation to achieve orgasm, whether by manipulation of the genitals or otherwise, is considered to be **masturbation**. Nocturnal emission is a misnomer since orgasm may be achieved in the daytime in both men and women and ejaculation occurs only in males. The term **sex dream** is more accurate but is used synonymously with "nocturnal emission" or other variations in phrasing. The terms are self-explanatory. The basic feature of a **homosexual** relation to achieve orgasm is that the partners are of the same sex, and the particular forms of physical activity are irrelevant to the definition. Similar means may be used by homosexual or by heterosexual couples to achieve arousal. Heterosexual intercourse includes penetration of the vagina by the penis (*intromission*) in the sexual outlet known as **coitus**. **Heterosexual petting** refers to any physical contact between partners of opposite sex in which sexual arousal is sought but without subsequent intromission. The category of **animal contacts** is defined according to the nature of the participants but, excepting anatomical limitations, there are similar physical activities to those that are undertaken by two human participants in eliciting orgasm.

The sections that follow are meant only to be illustrative of particular influences on sexual expressions, and they are not encyclopedic or complete. Some situations of a transient nature or those that may be imposed externally as in institutions of various sorts are excluded from the discussion. The major thrust of the relevant studies is to seek those biological and cultural characteristics that may show consistent relationships to rates and patterns of sexual expressions that culminate in orgasm.

Comparative Studies

Among the many lines of analysis reported by Kinsey and his associates were those in which particular characteristics were shown to influence patterns of sexual expression that were undertaken to achieve climax. The most important of these characteristics were the sex, marital status, age, social class, age at puberty, religion, and urban versus rural background of the individual. We will consider only the influence of marital status and sex of the individual to illustrate particular patterns and their frequency in relation to these two variables. The data were assembled according to age groups in five-year intervals similar to those mentioned earlier for orgasmal weekly frequencies in males and females.

There was a striking difference in pattern and frequency of varied sexual outlets for single versus married women (Fig. 10.5). Married women exhibited essentially similar patterns of sexual experiences and in approximately similar proportions for modes of experience between the ages of 16 to 55. In each of the age groups there was no less than 80 percent of orgasms attained by coitus. Masturbation accounted for about 10 percent of orgasmic experiences in each age group, and there were minor or negligible frequencies for the other four modes. Single women, on the other hand, achieved orgasm primarily by masturbation until the age of 21 after which the frequency was reduced to between 40 and 50 percent, with a steady rise toward the higher percentage from 41 to 50 years of age. Coitus was the second most frequent means of reaching climax, reaching a peak value of 43 percent of total outlets for the 41–45 age group. There was a steadily increasing frequency of coitus from postpubescence years up to the age of 30, then a relatively sustained frequency of about 30 percent until 40 years of age after which the experience of coitus increased to 40 percent or more for the women between 41 and 50 years of age. Homosexual relations accounted for steadily increasing proportions of orgasms up to the 36–40 age group, after which there was a sharp decrease to about 5 percent in women between 41 and 50 years of age. Heterosexual petting occurred most frequently between 16 and 25 years of age, accounting for about 20 percent of the total orgasms. Petting declined in frequency between 26 and 40 years and accounted for about 5 percent of climaxes in women be-

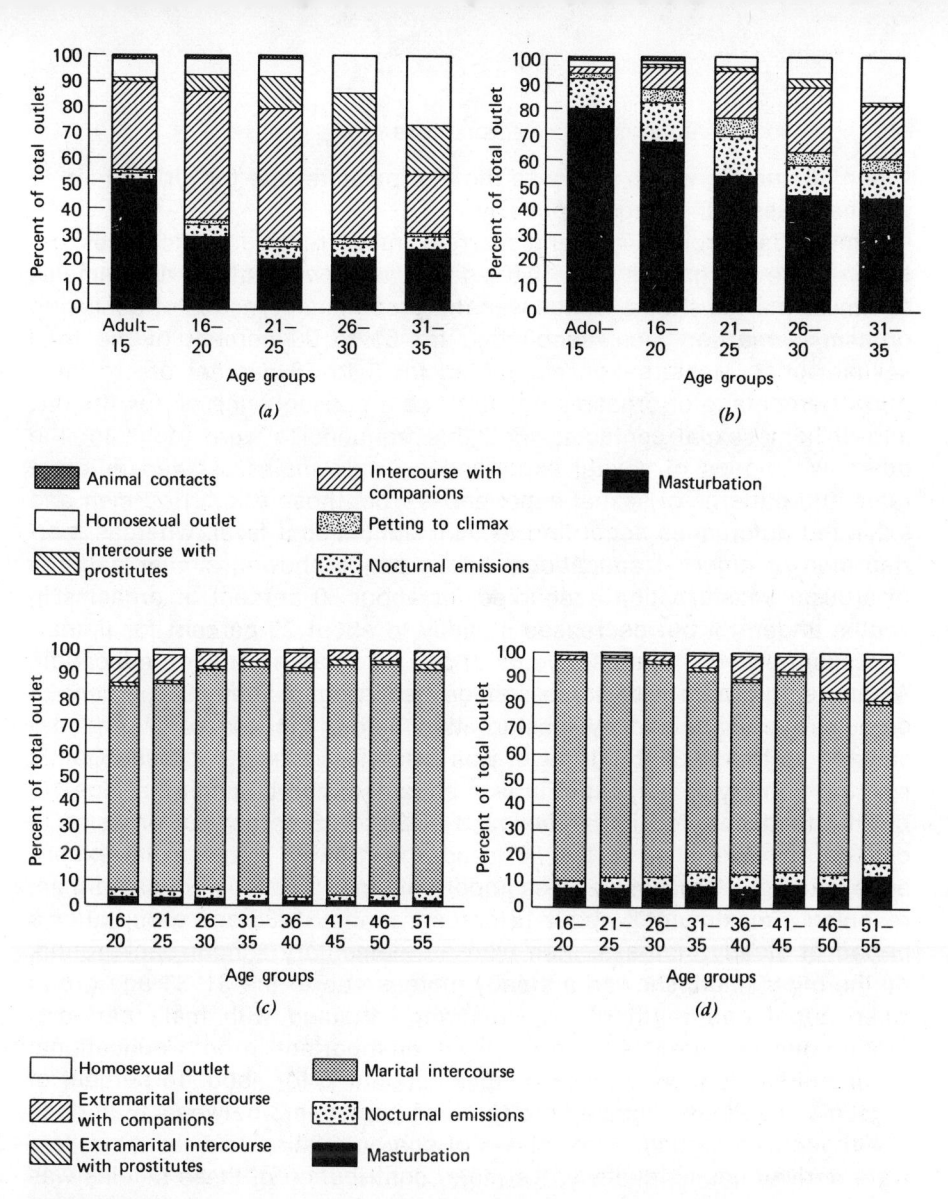

figure 10.6
Sources of orgasm expressed as percentage of total sexual outlet among males of different educational backgrounds and marital status, by age groups: (*a*) single males with elementary-school educations, (*b*) single males with some college education, (*c*) married males with elementary-school educations, and (*d*) married males with some college background. (From A. C. Kinsey and colleagues, *Sexual Behavior in the Human Male*, pp. 490–493. Philadelphia: Saunders, 1948. Reprinted with permission of the Institute for Sex Research, Inc.)

tween 36 and 50, which also was the approximate rate for girls under 15 who had passed the stage of puberty.

Similar data from single and married males revealed differences between these two groups (Fig. 10.6) as well as between males and females of similar marital status. Coitus was the predominant means of achieving orgasm in married men, accounting for 60 to 90 percent of the total sexual outlet. Masturbation accounted for 5 to 10 percent of orgasms; a low percentage of orgasms occurred as a consequence of sex dreams and of homosexual contacts; negligible frequencies were found for the other two modes of sexual experience. Single males showed different rates and patterns of sexual experiences from those of married men and exhibited differences according to their educational level, whereas married men of different educational backgrounds showed similar patterns of arousal. Masturbation accounted for about 50 percent of orgasms in youths under 16, but decreased steadily to about 25 percent for unmarried men between 16 and 35 years of age who had no college education. Among single men with some college background about 80 percent of orgasms were attained by masturbation before the age of 16 and decreased up to age 35, but it never was much less than 50 percent. Coitus was the primary sexual outlet in less educated men beginning at age 16, but it only accounted for a maximum of 20 to 25 percent of orgasms in college educated men in the 26 to 35 age groups. Homosexual experience leading to orgasm was less important among college educated men, reaching a maximum of about 15 percent in the 31–35 age group after a period of steady increase. Men with an elementary school background, on the other hand, showed a steady increase up to the 31–35 age group when about one-fourth of orgasms were attained with male partners. Petting and sex dreams were relatively unimportant in both educational groups of single men, but sex dreams accounted for about 15 percent of orgasms in college educated males in the peak years between 16 and 25.

Although no particular correlates of characteristics and sexual outlets were derived unambiguously, the major contribution of these studies was the revelation of a spectrum of variation in human sexual experiences. A great deal more must be known before it becomes possible to determine the relative inputs of biological, social, and psychological factors in the expressions of human sexual behaviors. Cross-cultural comparisons from data concerning about 200 different preliterate ("primitive") societies were made by Ford and Beach in 1951. Although the data were descriptive, not quantitative, they found that there was considerable variability among different peoples in the frequency of sexual experiences leading to climax. Sex taboos existed in every society in some form, as they do in modern cultures. The unitary basis for human sexual behavior is *variation*, which is predictable for a species that regulates its activities by complex genetic and cultural interacting systems.

Variations in Human Sexual Behavior

Any discussion of this topic leads to a decision on terminology before proceeding with the subject matter. The term "variation" implies an observation about which all or most people could agree; the term "deviation" implies a judgment with overtones of morality, legality, pathology, or similar connotation. This conflict in viewpoint usually does not involve aberrant behavior, such as psychotically motivated sexual murders. All judgments are essentially arbitrary, since gradations of behavioral modes clearly exist and various combinations of activities commonly are practiced by humans. Taking the predominant theme of human sexual behavior to be heterosexual intercourse, which is amply vindicated by quantitative data as well as descriptive observations, we may look systematically for patterns of variation in the major behavioral theme. Most discussions follow the general classification originally proposed by Freud in which there is a "sex object" and a "sexual aim" as parameters for comparison. The person of the opposite sex is the "object" and the wish for coitus is taken as the "aim" that serves to define the norms or standards of heterosexuality. Accordingly, one may pursue the subject systematically by characterizing variations in "sex object" and "sex aim" to discern alternative patterns that are necessary to attain a fuller understanding of human sexual behavior.

Variations in Object Choice

The alternative sex object to the heterosexual norm (a person of the opposite sex) may be a child, a close relative, a person of the same sex, an animal, an inanimate object, or a number of other possibilities. Just from this listing it is clear that we have not provided a full enough description of the standard heterosexual sex object, since we failed to include details of age and genetic relationship.

Homosexuality

The selection of an adult of the same sex as a partner in sexual acts is generally known as homosexuality. This term often is used pejoratively, so that some attempts have been made to use the term *homophile* as a suitable substitute. In addition to this merit, homophile could apply equally well to men and women, whereas *homosexual* generally signifies a male, and *lesbian* has been used to denote the female homophile. A recent term suggested by a more activist group is "gay," which also can describe either men or women who experience relations with an adult of the same sex.

The information that is available from several sources indicates that two to four percent of the adult males and one percent or fewer of adult females in the United States practice homosexuality exclusively. Similar figures have been cited from comparable estimates for countries such as Germany and Sweden. A much higher percentage of adults engage in homosexual experiences at one or more times in their lifetime, but do not do so exclusively. The concepts of masculinity and femininity are attributes that have been vaguely defined on cultural criteria. There is no demonstrable correlation between gender identity and sexual orientation, and no evidence exists from anatomy, chromosome analysis, or hormonal studies for biological differences between homosexuals and heterosexuals.

Homosexuality is not the dominant mode of sexual behavior in any known human or animal society, but it is tolerated or even incorporated into the social structure of many human cultures to varying degrees. Some cultures institutionalize homosexuality, designating particular individuals to assume the role of a member of the opposite sex. These instances usually are restricted for exploitation only by the more powerful decision-making members of the social group, and the practice is forbidden to all other members of such a society.

Some individuals are **bisexual**, engaging in both homosexual and heterosexual activities. **Transsexuals** are people (usually males) who prefer to be members of the opposite sex, but they are not homosexual in viewpoint, according to some psychiatrists. Sex transformation can be achieved by surgery that involves the removal of penis and testes and the creation of an artificial vagina by reconstruction of pelvic tissues. Such individuals are incapable of reproduction because the testes have been removed and no transplantation of ovaries and uterus is undertaken, but they may experience vaginal coitus.

Incest

Sexual relations between closely related individuals, not merely parents and offspring, is rare and is the one universal taboo in all currently known human societies. Although permitted in many religions, marriages between first cousins is specifically forbidden by the Roman Catholic church. Intermarriage between brothers and sisters was a permissible exception for members of some ancient royalties, as among the Egyptian pharaohs. Cleopatra was a product of such a sibling mating, and she herself practiced intercourse with her brother-husband, although no progeny resulted.

The virtual universality of the incest taboo has prompted various hypotheses that attempt to trace its origins to our ancient past. One suggestion has been that incest was forbidden originally because of its disruptive effect in family units of small communities in which early humans lived. Prevention of incest thus would have strengthened the bonds within the community and been of adaptive value in survival of the group. Another suggestion proposed that out-matings would have been actively encouraged, thus incest discouraged, leading to the enlargement of kinship groups to form more effective sizes of communal families. This also would have been adaptive and increased the probabilities for survival and well-being in small groups of humans. It also has been suggested that the deleterious genetic effects of inbreeding would have prompted the establishment of incest taboos. It probably was an influential factor in some cases, because numerous generations may have observed some level of deleterious effect even if the genetic basis was not understood.

Other Variations

An individual who practices **fetishism** is one who is aroused sexually by an inanimate object, usually clothing, as a consistent alternative to another person. Fetishism usually is incorporated into other sexual activities, especially masturbation. A **transvestite** practices a form of fetishism in that such an individual achieves sexual satisfaction by wearing the clothes of the opposite sex. Most transvestites are males, and such individuals may participate as females in some primitive societies. They often are considered to be "shamans," endowed with magical powers, and sometimes they may be married to the chief male of the group, but not to others in the society. Wearing women's clothing does not always indicate sexual variation, as in the case of the classical Kabuki theater of Japan, in which no women are permitted to perform and all the roles are played by males in these dramas.

According to the data presented in the Kinsey reports, 8 percent of adult males and 3 percent of adult females indicated that they had engaged in sexual activities with animals at one or more times. It was more frequent among people from rural areas, but all together only accounted for less than 1 percent of total sexual outlets.

Mythologies are filled with stories of interactions between people and animals, although the animal frequently was a deity or monarch in disguise. Symbolism was the clear intent in many instances, and not a literal situation. It is more difficult to interpret illustrations from past cultures as having symbolic versus a literal connotation.

In addition to the varieties of object choice described above, other manifestations also have been described. Interest in prepubescent boys or girls, group sex in which more than two individuals participate in a common sexual activity, and other variations are known.

Variations in Sexual Aim

There are a number of variations in which the sexual aim is to initiate arousal without intercourse. Observing other individuals having intercourse is a feature of **voyeurism**. Exposure of the genitalia to achieve sexual stimulation is called **exhibitionism** if the chosen witnesses are strangers and no intercourse is intended. Sadomasochism involves the elements of **sadism**, in which arousal is achieved by inflicting pain on another person, and of **masochism**, in which sexual gratification is achieved by receiving pain. Other variations, in addition to these, also have been described.

Two of the six principal modes described by Kinsey as significant mechanisms for attaining orgasm were outlets in which the activity was *solitary*. Sex dreams ("nocturnal emissions" or "wet dreams") were a minor component, varying from 2 to 8 percent in males and 2 to 3 percent in females of the total sex outlet of the sampled individuals. A large percentage of the individuals reported having sex dreams at some time in their lives, even though their frequency may have been relatively low. Almost every male had experienced orgasm through sex dreams at some time, and about two thirds of the females.

The second major form of **autoerotic activity**, masturbation, is usually dependent on deliberate acts of self-arousal, whereas sex dreams are not. As discussed earlier in this chapter, 92 percent of males and 58 percent of females were reported by Kinsey and his associates to have masturbated to climax at some time during their lives. There were a number of correlates that were significant in relation to frequency differences for this particular sexual outlet, especially the sex of the individual, age, educational level, and marital status. Similar cultural influences apparently contribute to the frequency of masturbation in different human societies, but the practice is essentially universal among human social groups. Subhuman primates have been observed to masturbate both in free-living and in zoo populations, the practice occurring more frequently among male than female members of the various monkey and ape species. Autostimulation apparently is rare or lacking in nonprimate mammals, although scattered observations have been reported in the literature.

Diseases Involving Reproductive Organs

Of the several categories of diseases, two are more important and will be discussed briefly. The **venereal diseases** are primarily transmitted through sexual intercourse. They are *infectious*, since there is a known microorganism that causes the disease, and they are *contagious*, since they may be transmitted directly from one infected person to another. Not all infectious diseases are contagious; for example, malaria infections are initiated only by transmission of the causative protozoan from mosquitoes to the host individual and not from one sick person to another. The second major category of afflictions are **cancers**, and they affect the breast, cervix, prostate gland, and penis more often than other accessory organs associated with human sexual activities.

Venereal Diseases

Gonorrhea

The infection is caused by a bacterium called *Neisseria gonorrhoeae*, which appears as pairs of spherical cells when viewed through the microscope. From available data it seems that infection primarily is initiated when the bacteria enter the body through the mucous membranes of the genitourinary tract, and not by vicarious contact with other parts of the body or contaminated objects. The primary symptom is a purulent urethral or vaginal discharge which is the result of an acute inflammatory response in the earliest stages. The discharge occurs more consistently in males, and sometimes is absent or negligible in females in the 3 to 10 days between infection and symptom display. The inflammation may subside in a few weeks, become chronic, or spread to other internal organs. The development of scar tissue in the Fallopian tubes after gonorrheal infection often is one cause of sterility in females, since ovum passage may become blocked. The compulsory practice of putting silver nitrate drops in newborn babies' eyes stems from the previously high rate of incidence of a form of blindness which developed in children who contracted gonorrheal infection of the cornea during passage through an infected birth canal at delivery.

Gonorrhea responds relatively quickly to antibiotic therapy, although higher concentrations of the drugs may be necessary since numerous bacterial strains have appeared with increased levels of resistance to antibiotics and other chemicals. There is little or no immunity developed by individuals to subsequent infections, so that treatment is required each time a new infection has been initiated. Approximately 90 percent of cases respond rapidly (within about 12 hours) to the antibiotic if treatment is given promptly. Infections usually can be prevented in several ways: use of a condom during intercourse, prompt and efficient washing after intercourse, or antibiotic treatment within a few hours after exposure.

Syphilis

Unlike gonorrhea, which apparently was well known in ancient cultures according to Chinese, Egyptian, Greek, and other manuscripts, syphilis may have originated in the New World and spread to other regions after 1492. The disease was not described in Europe until after Columbus and his crew returned from their voyages, and it appears to have been carried to Asia from Europe by an assortment of European explorers and navigators beginning with Vasco da Gama in 1498. Clinical diagnosis of pre-Columbian Indian bones has provided evidence that syphilitic infections probably did exist in the New World before European visitations.

The syphilis infection is caused by a spiral-shaped bacterium called *Treponema pallidum* and appears two to four weeks later as a painless, hard ulcer (called a *chancre*) at the site of contact with the bacteria. The appearance of the chancre on external genitalia, mouth, skin, or on vaginal or cervical areas constitutes the *primary stage* of the disease. This skin lesion generally disappears a few weeks after its formation, and the secondary stage of the disease is manifested several weeks afterward. The *secondary stage* produces a set of vague and general symptoms, difficult to associate with the earlier chancre stage: general rash on the skin and assorted aches and pains. A latent stage occurs shortly afterward that has a variable duration; the bacteria settle into bones, blood vessels, central nervous system, and other tissues. About 50 percent of untreated cases exhibit *tertiary stage* symptoms even years later, these including blindness, deafness, heart failure, loss of muscular control, and other gross defects. The disease is fatal, although antibiotic therapy still may be useful if no vital organ has been seriously damaged during this tertiary stage. About nine out of ten children of syphilitic mothers either miscarry, are stillborn, or are born with this disease. Children with congenital (present at birth) syphilis show impairments of sight, hearing, bones, and other organs, but antibiotic therapy may alleviate many of these manifestations if treatment is begun during early infancy.

Cancers of Reproductive Organ Systems

The reproductive system is more prone to cancer than any other parts of the body. Of the more common occurrences, two appear in women and two in men.

Cancers in Women

About 5 percent of all women ultimately develop **cancer of the breast**, but its incidence is rare in women under 25 years of age. Breast cancers respond to sex hormones, which explains in part the spread of this cancer under the conditions of high estrogen levels during pregnancy. The oral contraceptive pill is regarded with some caution by many physicians and biologists because its estrogen content may lead to malignancy. We will not know if this is true until at least 1980 when sufficient time has elapsed for thorough follow-up of women who have taken the pill for 20 years.

Early diagnosis and treatment permits a high rate of survival of cases, since the cancerous tissues are removed before *metastasis*, or spread of the cancer to new locations and development in those sites. About one third of the cases survive ten years or longer after removal of the entire breast or some part of it. Reduction of the estrogen level, by removal of the ovaries, often is beneficial in treatment of breast cancer. Unless examinations are conducted regularly by the woman or her physician, the painless lump in the breast could go unnoticed for a long time and perhaps spread to other areas.

Cancer of the cervix is the second most common malignancy in women, occurring at some time in about 2 percent of all women. The incidence increases steadily from a rare frequency before age 20 to higher levels, and a maximum at 45 years of age, after which the frequency declines. Unlike cancer of the breast, cervical cancer is more common in women who have been sexually active and have borne children. Also, there may be no particular symptoms for the first five or ten years unless a **Pap smear test**, developed by Papanicoulou, indicates the presence of cancerous or incipiently cancerous cells when the specimen is viewed with the microscope. Treatment during this early stage is extremely successful, coming close to 100 percent.

The survival rate after treatment for localized cervical cancer is about 80 percent after five years, but drops to a very low value if the disease has spread to other pelvic organs before treatment. The ten-year survival rates are lower, with about one fourth of all women still alive after treatment of the cancer in all stages, including postmetastasis.

Cancers in Men

Cancer of the **prostate gland** is the commonest form of cancer in men, but rarely occurs before age 50 and is unlikely before age 60. About 70 to 80 percent of cases are found in men between the ages of 60 and 80, and one fourth of men in their 90s develop cancer of the prostate. The mortality rate is rather low, however, because the growth occurs slowly and usually does not spread to other organs. Older men with such cancers are more likely to die of other causes than from this disease directly.

A common surgical treatment is removal of the testes, since androgens stimulate the cancerous growth. Similarly, therapy usually includes injections of exogenous estrogens, since these hormones are antagonistic to the cancer and slow its growth. Early detection and treatment usually are beneficial.

Cancer of the **penis** accounts for about 2 percent of all cancers in male Americans, so that it is relatively rare in the United States. It may be quite common in some countries, but occurs more rarely in men whose religion includes circumcision, such as Jews and Moslems. It is not known why this should be the case, but these malignant growths usually occur at the tip of the penis, and it may be that removal of the foreskin reduces the chances for accumulations of chemical or viral agents that may contribute to the development of malignancy.

Suggested Readings

Beach, F. A. (ed.), *Sex and Behavior*. New York: John Wiley & Sons, 1965.

Benjamin, H., and I. B. Pauly, The maturing science of sex reassignment. *Saturday Review* (October 1969), 72.

Etzioni, A., Sex control, science, and society. *Science, 161* (1968), 1107.

Ford, C. S., and F. A. Beach, *Patterns of Sexual Behavior*. New York: Harper & Row, 1951.

Katchadourian, H. A., and D. T. Lunde, *Fundamentals of Human Sexuality*. New York: Holt, Rinehart, Winston, 1972, pp. 149–318.

Kinsey, A. C., W. B. Pomeroy, and C. E. Martin, *Sexual Behavior in the Human Male*. Philadelphia: Saunders, 1948.

Kinsey, A. C., W. B. Pomeroy, C. E. Martin, and P. H. Gebhard, *Sexual Behavior in the Human Female*. Philadelphia: Saunders, 1953.

Lieberman, B. (ed.), *Human Sexual Behavior: A Book of Readings*. New York: John Wiley & Sons, 1971.

Masters, W. H., and V. E. Johnson, *Human Sexual Response*. Boston: Little, Brown & Co., 1966.

Masters, W. H., and V. E. Johnson, *Human Sexual Inadequacy*. Boston: Little, Brown & Co., 1970.

McCary, J. L., *Human Sexuality*, 2nd edition. New York: Van Nostrand Reinhold, 1973.

Sherfey, M. J., *The Nature and Evolution of Female Sexuality*. New York: Random House, 1972.

Wallace, B., *Essays in Social Biology*, Vol. III. Englewood Cliffs, N.J.: Prentice-Hall, 1972.

chapter eleven

Sex Roles in Society

There is considerable interest today in the questions concerning sex role assignments in modern human societies and in a comprehension of gender identity in relation to human behaviors. The movements centering around women's rights and the rights of every human being to participate in social, sexual, and economic life as one chooses and as one's potentiality permits are of central concern to us all. In this chapter we will focus primarily on the interactions of biological and cultural determinants in producing the patterns of male and female roles in the human society and the family. There is little agreement among the different camps of opinion, and a rather casual approach to the interpretations of the available data. The controversies are maintained largely because the little information that we do have has been manipulated to conform to one set of conclusions or another instead of being assimilated into a comprehensive framework by which one may proceed to phrase pertinent questions and to direct the approaches that can provide appropriate answers and interpretations.

The central theme common to many studies and approaches concerns the relative levels of distinctness of male and female capabilities in fulfilling social, economic, and sexual roles in the amalgam of life activities. How much of the present levels of differentiation can we attribute to the genetic endowment and how much to the processes of societal conditioning and acculturation? What segments of the sex role and its concomitants are influenced by hormonal physiology, learning, and sexual morphology? It is impossible to discuss such complicated problems in a complete fashion in a brief chapter, but we can examine some aspects of the phenomena. To begin such a discussion we should examine some of the fundamental properties and the history of the modern human being.

Human Societies

Pattern Variability

The immediately obvious feature of social groupings in our species is its incredible *diversity*. There certainly are varying levels of affluence in the economy; the size of the aggregation of people varies from about 20 to many millions; the social group may consist of nomads or settlements of different sizes; fisherfolk, game hunters, agriculturists, foragers, or urbanites; child care may be vested in community centers or exclusively in the home, or some combination of these modes; governance may be by one or a few individuals or by the entire community; and other features too numerous to list. This diversity absolutely does not exist in any other species. No matter what group one may study among wolves, chimpanzees, monkeys, lions, or elephants, there is a relatively limited format of social organization and economy throughout each species. Some variability has been observed among different populations of a single species, but not one of these began to approach the profound diversity of human society.

Another characteristic of the human species that is unique is the capacity for *flexible and rapid change* in social behaviors. The introduction of agriculture 10,000 years ago contributed to the development of urbanized living, and in some countries as many as 70 percent of a once predominantly rural, farming population now lives in cities. Numerous peoples of the islands of the South Pacific underwent a change from a formerly "Stone Age" culture to one more typical of twentieth-century civilizations within the brief time span of one or two generations. The opening of Japan to western influences in the midnineteenth century led to a series of rapid modifications in the economy and social structures of that country, and we now consider Japan to be the most "westernized" of the Asiatic countries. Such changes as have been recorded to occur in some primate societies are of a relatively trivial dimension in comparison with those in human groups. A habit of washing chunks of sweet potatoes before eating was observed to occur in a colony of Japanese monkeys, and the innovation spread by imitation to a substantial part of the colony in a brief time. Chimpanzees primarily are vegetarian foragers for food, but they may aggregate temporarily into small bands to hunt for small game animals. These hunting episodes occur intermittently and do not alter the primary behavior of foraging for foods in the surrounding vegetation. Although the chimpanzees obviously relish these occasional meat dishes, there is no concerted effort to organize and plan for these to become a proportionately greater percentage of their diet.

The basic human social group includes adult males and females along with adolescents, juveniles, and infants that belong to the adults of the community of individuals. We thus resemble other anthropoid primates in being social animals grouped together in intraspecific aggregates that exhibit effective interactions in social and sexual behaviors. Except for the social lemurs of Madagascar, other prosimian primates do not demonstrate social group organizations. Sociality would thus appear to be an ancient heritage that extends back in time to the origin of the anthropoid branch of the primate order of mammals.

There usually are more females than males in primate societies, whereas most human social groups include approximately equal numbers of men and women. There are variations on this theme, too, and some groups include more members of one sex than the other. The disparities in the percentage of males and females sometimes lead to differing patterns of sexual and reproductive behaviors, as in societies that are *polygynous* (several women per man) or less often *polyandrous* (several men per woman) in family structure. The predominant family pattern in human societies is *monogamy* (one marriage per person). Gibbons also tend to be monogamous, but these hominoid primates generally live in small nuclear family groups comprised of one male, one female, and their young instead of in a community of families.

Care of the Young

The commonest pattern of parental-filial interaction in human societies is one in which the immediate parents care for their own offspring. In some aboriginal or preliterate human societies the care of the young is shared by a consortium of adults or by particular individuals whose responsibilities include care of adolescent members of the community. In the latter cases the juveniles are separated according to sex, and girls are tutored by a different set of adults than those who house and teach young boys of the community. In an increasing number of western countries the care of the infant and young child soon is taken over by community centers, after which the child is cared for predominantly by the educational system. There is no loss of identity between parents and child, but the education and provision responsibilities are monitored by unrelated adults of the social group. This pattern presumably would become more widespread in the United States if there were sufficient day-care centers available such as those in Scandinavian and Soviet countries, China, Israel, and other states in which most of the adult men and women are working members of the economy. This variability and flexibility of pattern may be observed among different species of primates, but never within a single species as is the case with human beings.

Although we consider the care of infants and young children to be of paramount importance today and as palpable signs of our "humanity," there are societies in which children virtually are discarded by the parents and left to fend for themselves at a very tender age. A recent study of the Ik people of east Africa has created shudders among readers of the reports describing cruelties and afflictions that the Ik visit on one another and on their children. These people were displaced from their tribal hunting grounds about 50 years ago when a national game preserve was established. Barred from their usual sources of food and unable to exploit the barren hills in farming, the Ik have deteriorated socially as they slowly starve to death in a changed economy and ecology. Economic factors also contribute to incidents of infanticide in some preliterate human societies, as they did to the high rates of abandonment and death of infants that characterized western European countries during the eighteenth and early nineteenth centuries.

The Big Hunting Mutation

A popular theme in current literary circles has been that of inherent male supremacy, advocated by a small and dedicated group of professional and amateur anthropologists. These men have proposed that the current behavioral modes of the human species represent a genetic legacy from our hunting ancestry, which extends at least 15 million years back in evolutionary history. According to these ideas, the social and sexual behaviors of modern human beings have evolved in relation to the acquisition of hunting in place of foraging and food gathering as the predominant life-style of our ancient ancestors. As hunters, our forefathers came to assume a greater share of the burden of ensuring the survival of the species, while our foremothers assumed a more and more subordinate and trivial role in the directionality of evolutionary improvements.

Men were the hunters, and their success depended on the increasing development of skills, strengths, fleetness, intelligence, and cooperativeness. Lacking anatomical weapons such as great canine teeth, sharp claws, huge bulk, and muscle masses, the prehuman male could persist only by incorporating brains for brawn and homemade weapons for his inadequate anatomy. To succeed in the hunt for large game animals, groups of men had to join together in the hunt, and this could be accomplished most effectively if they cooperated and sublimated their inherent aggressiveness toward one another. Cooperation requires more intelligence and organizational skill, thus providing further selection pressures that would lead to adaptive increases in brain size and complexity in the prehuman males. Needless to say, what little benefits accrued to the females of the species were attained only because they had access to the gene pool that was continuously upgraded by mutations leading to improvements in males. Speed, cleverness, skills, organizational proclivity, and other desirable attributes for the hunting males were of no consequence in the daily bustlings around the campfire for women who merely bore children, cared for the children, educated the children, and accomplished the impoverished variety of domestic chores. Except for the fact that only women could conceive and bear offspring, they really had little purpose in the hunting economy, unless we also include sex in the array of reproductive behaviors.

As noted by Desmond Morris, "man is the sexiest primate alive." Much of the argument and explanation presented in favor of male superiority by Morris and other authors of similar outlook has centered around attempts to interpret human sexuality in terms of our hunting ancestry. There are some indisputable observations, such as the lack of estrus in human females, but the interpretation of these evolutionary happenings has been subjective to an exceptional degree. The usual narrative contends that prehuman and human sexual behavior and the differentiation of societal roles are a direct outcome of a hunting ecology and economy. Men away on the hunt might be gone for days or weeks, and it would be extremely awkward for the species if the female ovulated and became sexually receptive while the males were not at home. It would be a clear benefit to the species if genetic changes that led to a loss of estrus were to become incorporated into the species, once such mutations arose by chance. The female then would be sexually receptive at any time and thus would be available whenever the hunters were at home. But this in turn provoked new problems, since males away on the hunt might be "worried" about their sexually receptive females who remained in the camp. The development of *pair bonds* clearly would solve this new problem. Monogamy thus would come to characterize prehuman and human hunting societies. The female pair-bonded to a particular male would be preserved for him whenever he had the time to be at the now-stabilized home and not on the savannah looking for the food that ensured the survival of the entire species.

According to the proponents of male superiority we thus are a consequence of the steady incorporation of genes that conferred particular adaptive features to a species devoted to hunting big game. The intelligence and skill in organizational endeavors, whether in government or industry or business, is the genetic legacy of the male handed down from generations of male ancestors whose talents were honed by natural selection acting on the "bio-grammar," or encoded information in our DNA that directs these behavioral differences between men and women to the present day.

There is no question that men and women differ in a number of characteristics. The most obvious differences are anatomical traits that develop as a consequence of interacting genic and environmental influences that modulate the neuroendocrine systems and contribute to behavioral development. There is more muscle mass in men and more fat deposits in women; breast development is hormonally enhanced in women and not in men; hair is distributed equally over the body in men and women, but in males there is differential hair growth in more areas of the skin than occurs in females. Gonadal products are different in the two sexes; women menstruate, gestate, and lactate, whereas men do not; sperm are produced daily in the hundreds of millions, but only one ovum is released per month in females; procreation ceases in women earlier than in men. All these disparities are responsible in some measure for eliciting behavioral differences between the sexes. These and other primary and secondary sex traits are different for males and females, but there are no demonstrable differences in intelligence or organizational talents; in literary, artistic, or scholarly skills; or even in the quality of parental love and affection. Although men generally are physically stronger, there often is particular effort directed toward developing such strength in males, while equivalent training is discouraged for females in most societies. For most of the human traits there is as much difference between selected pairs of individuals of the same sex as between selected individuals of the opposite sex in relation to ability to function and contribute effectively in human societies.

Some features are not absolutes and are not inherent in development of each individual of a particular sex. The average height of American men is greater than the average height of American women, but American women generally are taller than males of many other countries, such as those of southeast Asia. Among other primate species the male in *any* group of the same species may be twice the size of the female, but this is not true for humans. Among other species the behavioral patterns are relatively similar from one social group to another within a species, but this is not true for humans. The general trend in human evolution has been toward lesser sexual dimorphism. Indeed, we artificially emphasize or underline the gender of the individual in our clothed societies, presumably because of the importance of maintaining some set of signals that identifies an individual as male or female. Young people may dress alike and let their hair grow long, but males concurrently have adopted styles of facial hair growth that signal their gender to the casual observer. We begin the tradition of emphasizing male and female as different from the time the infant is wrapped in a blue or a pink blanket. Each of us is conditioned to respond differently to males and females of any age, and elements of clothing or personal style are designed to provide the appropriate signals that can be understood by members of the society so that appropriate responses can be elicited.

In general, human beings are considerably more variable physically, socially, and sexually than any other species of life. There are major behavioral themes, however, and their apparent universality has been cited by some groups in support of their contention that sex role assignments reflect inherent instead of acquired qualities for males and females of our species. Whether one emphasizes the variability or the common themes, there still remain the major questions concerning the origins and traditions of sex role differentiation in human societies. We will next consider the major lines of evidence that underlie the interpretations and conclusions reached by proponents and opponents of the hunting origins of sex role differences in modern human social groups.

Principal Lines of Evidence

Three major kinds of evidence have been cited in relation to the concept of the prehuman hunter: the fossil record, comparative study of subhuman primate species, and cross-cultural patterns in preliterate human societies. Although each of the opposing camps may cite the same general kinds of evidence, the different interpretations are due to the nature or selection of the particular evidence described and the different assumptions which underlie the eventual interpretations or conclusions. This is not an unusual situation in scholarship, but it has been introduced to an astonishing degree in the discussions of superiority versus equality of individual human beings.

The Fossil Record

The primate order of mammals first appears in the fossil record dated about 60 million years old (Fig. 11.1). The most ancient members of the group were simple **prosimians**, of which a few types still exist in Asia (tree shrews, lorises, and tarsiers), the African continent (galago, potto, and bush baby), and in the Malagasay Republic on the island of Madagascar off the southeast coast of Africa (true lemurs and aye aye). At least 40 million years ago there already were members of the other

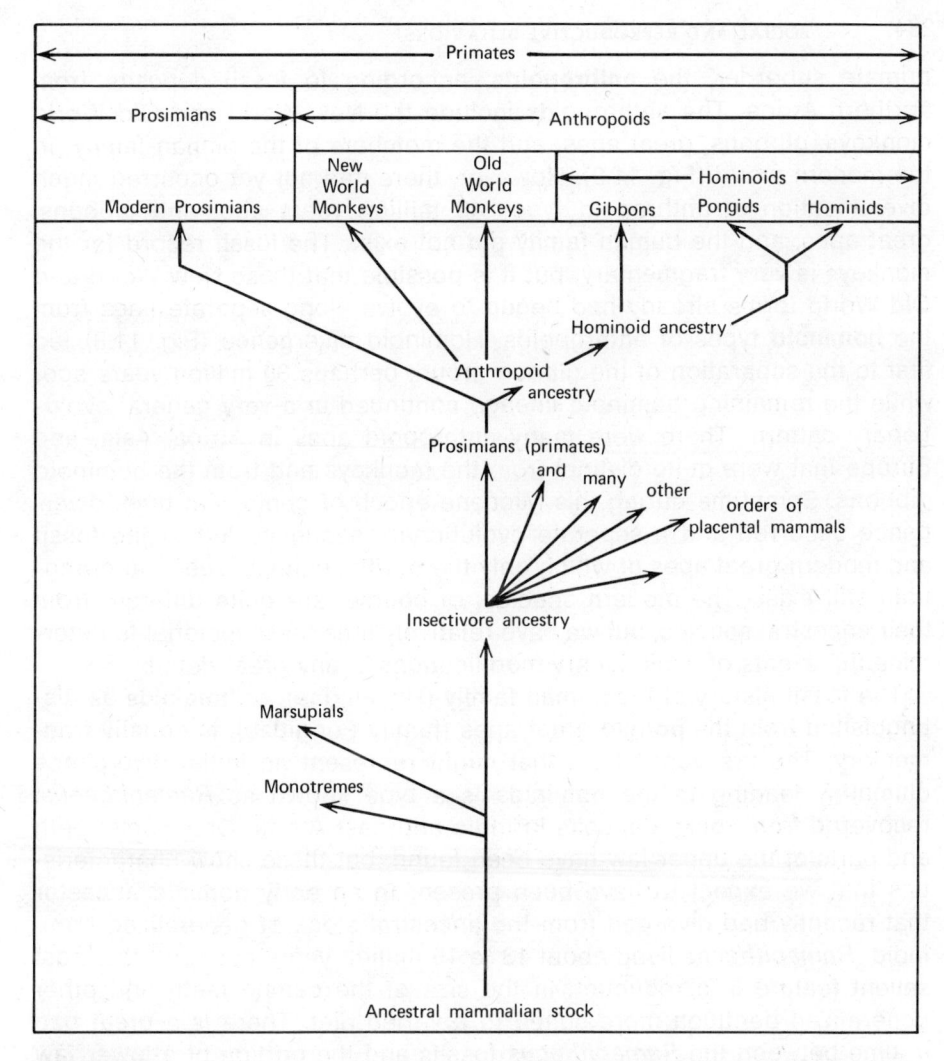

figure 11.1
Derivations and relationships of the primates, an order of placental mammals. The two major subdivisions include the more ancient prosimian lineage and the subsequent anthropoids. Anthropoid evolution led to the monkeys of the New World and the Old World and to a generalized hominoid line. Hominoid divergence first produced the gibbons and a later branching led to the pongids (great apes) and the hominids (human family).

primate suborder, the **anthropoids**, according to fossil deposits from northern Africa. The anthropoids include the New World and Old World monkeys, gibbons, great apes, and the members of the human family, in the modern world (Fig. 11.2). However, there had not yet occurred much diversification of anthropoid types 40 million years ago, and gibbons, great apes, and the human family did not exist. The fossil record for the monkeys is very fragmentary, but it is possible that these New World and Old World forms already had begun to evolve along separate lines from the **hominoid** types of anthropoids. Hominoid divergence (Fig. 11.3) led first to the separation of the gibbon group, perhaps 30 million years ago, while the remaining hominoid lineage continued in a very general evolutionary pattern. There were many anthropoid apes in Africa, Asia, and Europe that were quite distinct from the monkeys and from the hominoid gibbons. Sometime during this Miocene epoch of geological time, divergence occurred and a separate evolutionary sequence led to the fossil and modern great apes of which only the gorilla, chimpanzee, and orangutan still exist. The modern species, of course, are quite different from their ancestral species, but we have relatively little fossil material to determine the events of evolutionary modifications in any great detail.

The fossil history of the human family (Hominidae), or **hominids** as distinguished from the **pongid** great apes (family Pongidae), is equally fragmentary. The first fossil form that might represent an initial divergence ultimately leading to the hominids is a type known as *Ramapithecus*, recovered from rocky deposits in India and east Africa. Only some teeth and parts of the upper jaw have been found, but these show characteristics that we expect to have been present in an early hominid ancestor that recently had diverged from the ancestral stock of generalized hominoid. *Ramapithecus* lived about 13 to 15 million years ago, and the most salient feature is a reduction in the size of the canine teeth and other generalized dentition more suited to a varied diet. There is a great gap in time between the *Ramapithecus* fossils and the portion of a lower jaw dated to be 5½ million years old that was found recently in Kenya near Lake Rudolf. Two other important finds in this area are a 4 million-year-old elbow and a group of skull fragments plus leg bones from two other individuals, estimated to be 2½ million years old. These Kenya fossils are believed to be true members of the hominid family, but their precise relationships are somewhat controversial at present. It is not known for certain, until the material can be examined more intensively, if the youngest Kenya specimens are members of the **australopithecine** or **hominine** subfamilies of the Hominidae.

All members of the Hominidae are characterized as being exclusively bipedal species that possess upright posture, a larger brain, and generalized and varied kinds of teeth, as distinguished from the pongids,

figure 11.2
Some representatives of present-day primates. (*a*) lemur, (*b*) tarsier, (*c*) Rhesus monkey of the Old World, (*d*) gibbon, and (*e*) gorilla.

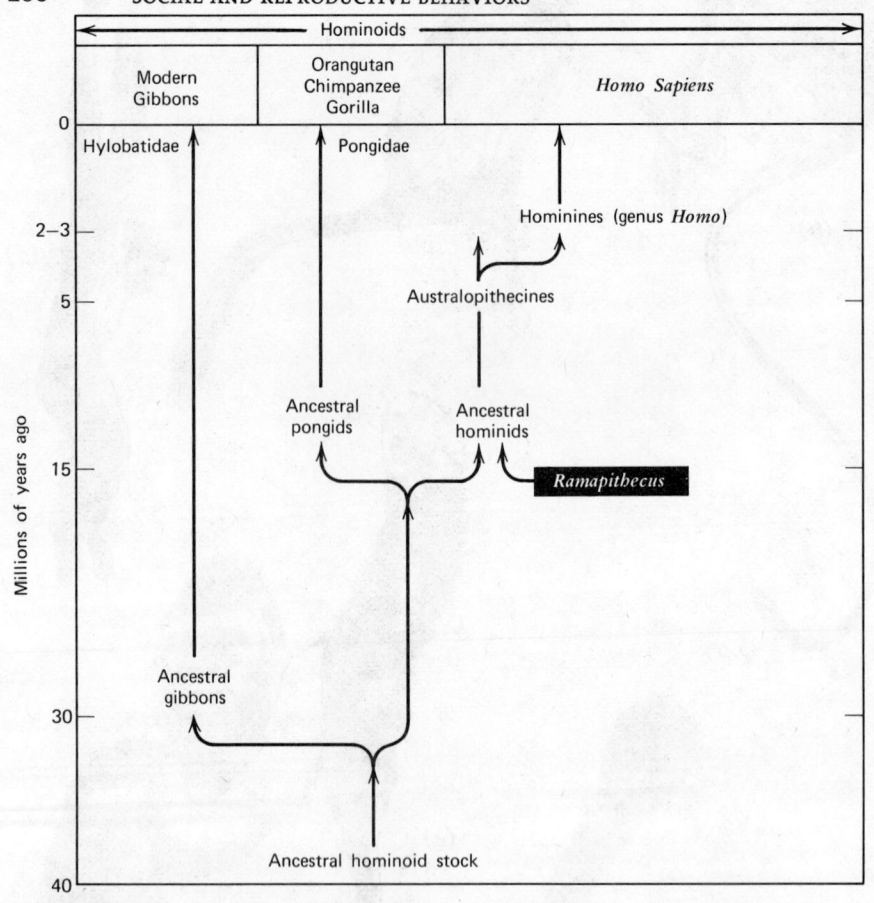

figure 11.3
Derivations and relationships of the hominoid group of the anthropoid primates. The gibbons diverged early, and a later divergence led to separate evolutionary pathways for the great apes of the pongid family and the hominid family to which modern humans belong. The australopithecines are believed to represent the more ancient hominids, and all are now extinct. Subsequent divergence from an australopithecine stock produced the hominines, all belonging to species of the genus *Homo*. The only remaining species of the family Hominidae is our own, *Homo sapiens*.

which use all four limbs for locomotion, have a relatively smaller brain in relation to body size, and possess certain dental traits, which include large and prominent canine teeth, plus other species characteristics (Fig. 11.4). From the fossil evidence it seems clear that the first significant modification in the ancestry leading to hominids was the development of *upright bipedalism* and a *generalized dentition*. The enlargement of the

Sex Roles in Society   237

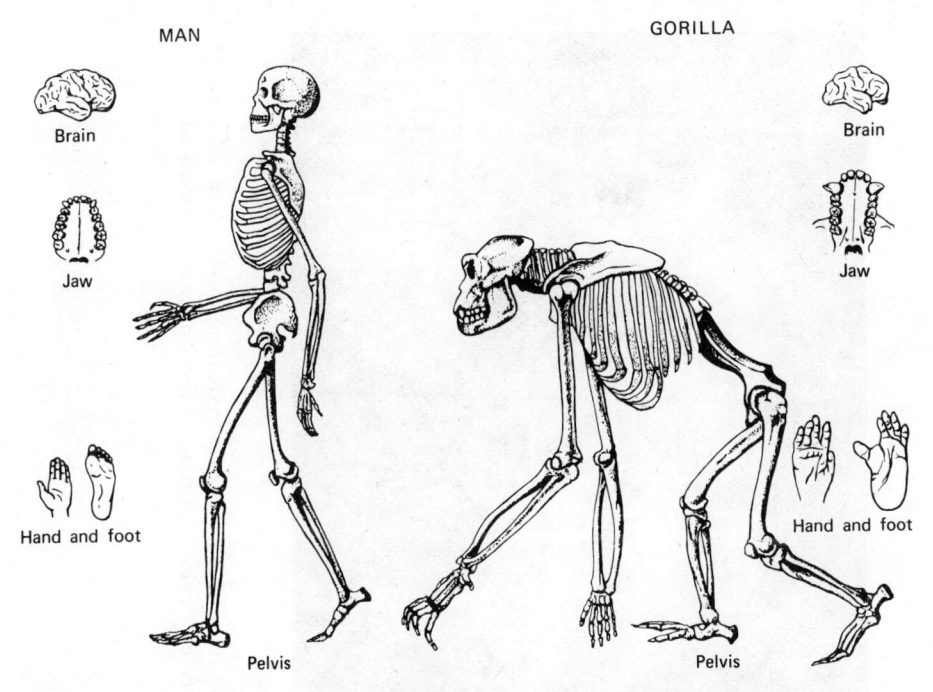

figure 11.4
A comparison of some structural differences between a gorilla (Pongidae) and a human (Hominidae), illustrating some of the significant distinguishing features.

brain was a later evolutionary development. The two kinds of hominids, australopithecines and hominines, are distinguished on the basis of tool-making abilities and relative brain size, with increase in the brain presumably indicating greater intelligence and the beginnings of the capacity for establishing tool-making cultures. All hominines have been classified as different species of the genus *Homo*, and the only surviving species is our own, *Homo sapiens*. At least one other fossil species of hominine is recognized by all, namely, *Homo erectus*, who existed for half a million years on the continents of Africa, Asia, and Europe, but became extinct about 200,000 years ago (Fig. 11.5). An earlier species, *Homo habilis*, has been identified by Louis and Mary Leakey from east African deposits approximately 1½ to 2 million years old, but there is no consensus among paleontologists and anthropologists about its specific identification and niche in hominid lineage. Neanderthal fossils also belong to the genus *Homo*, but their precise affiliation with *Homo sapiens* or their separate specific status is uncertain at present. These forms disappeared

figure 11.5

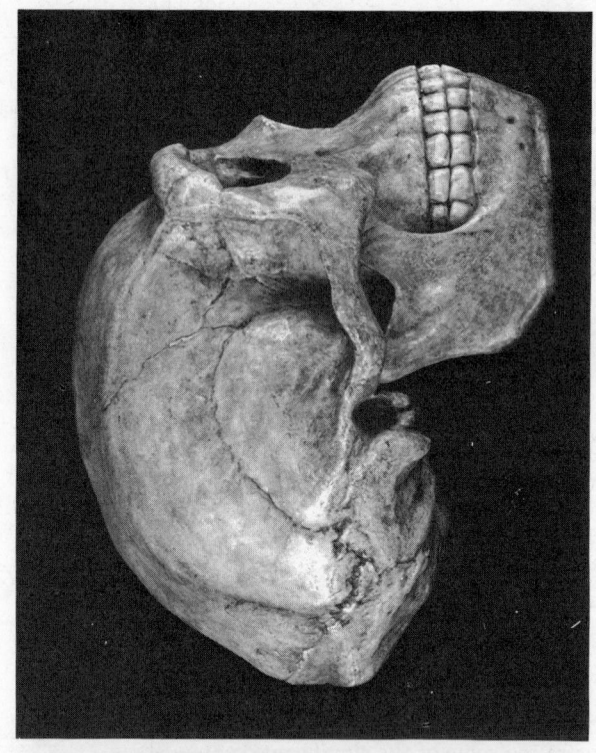

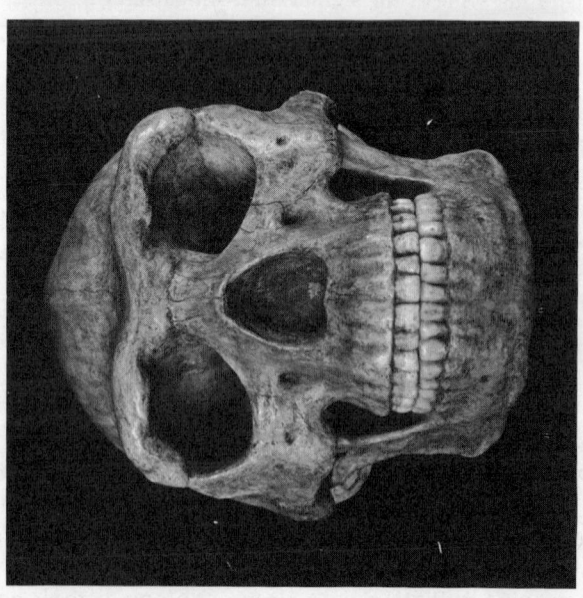

Front and side views of a reconstructed skull of *Homo erectus* from China. Note the retreating forehead, massive bony ridges over the eyes, large teeth, massive jaws, and the absence of a chin. (Courtesy of the American Museum of Natural History, New York.)

rather suddenly about 50,000 years ago, at about the time that modern *Homo sapiens* first appeared in the fossil record.

Interpretations of the Fossil Record. The proponents of the hunter theory to explain male superiority and the uniqueness of human social and sexual behaviors trace our conversion from foraging woodland vegetarians to hunters on the open savannah beginning with *Ramapithecus*. His altered dentition indicates a change in diet that could include meat and a reduction in the defense apparatus of the canine teeth that could imply that alternative modes of defense were employed. As a resident of the open savannah, *Ramapithecus* was surrounded by predators that could outrun him and that certainly were much more powerfully built. The reasoning is that this species only could have survived if he was a hunter in the harshness of the savannah, if he could defend himself against predators, and if he could outwit his pursuers and competitors. Selection pressures would have led to adaptive changes in the gene pool, such as increased intelligence and an improved locomotion for the open spaces by virtue of more and more bipedal activities. The modifications that first appeared in *Ramapithecus* presumably became enhanced during subsequent evolution and finally emerged in the hominines, which were more intelligent, exclusively upright and bipedal, and makers of tools that included weapons for hunting and defense.

There is no way to evaluate these suggestions, because the fossil record is so deficient, but some questions certainly can be raised about these interpretations. First, there is absolutely no evidence of tool making until about 2 million years ago, so that the supposition about effective hunting and defense tools for *Ramapithecus* and 10 million or more years of subsequent populations is entirely speculative. Second, the development of bipedalism as a means of locomotion is more related to an ability to walk long distances less tiringly than to speed in running away from pursuers and toward prey. The particular modifications of the hominid foot also are geared to efficient striding locomotion and are poorly suited to running. The fastest residents of the savannahs are four-legged, not two-legged; their running ability is enhanced by feet that place the toes on the ground, and not by a more supportive structure like the human foot; and the most successful hunters of the savannah are the lions and leopards, which also are powerfully built as well as fast. Third, primate species in which there is a primary role in defense exercised by males usually show exaggerated sexual dimorphism in canine tooth displays as well as in general body build and size (Fig. 11.6), yet the trend toward hominization reveals the opposite track toward decreasing sexual dimorphism. Fourth, residence in the savannah as opposed to open woodlands or deep forests has not led to divergence of hunting and foraging primate species. Baboons of the savannahs and gorillas of the forests

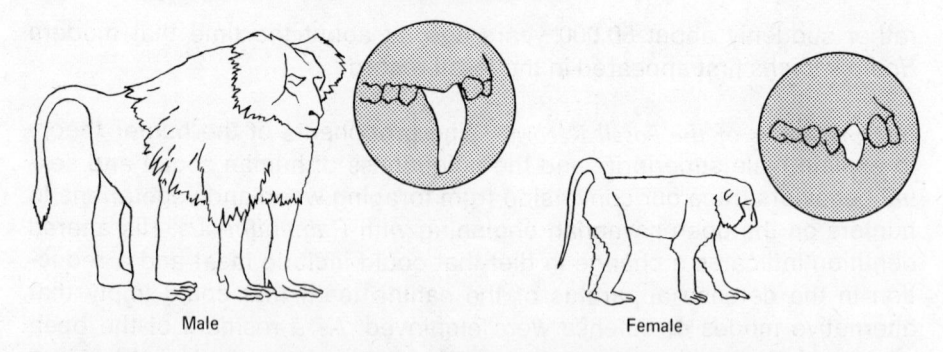

figure 11.6
Sexual dimorphism in the baboon. The male is about twice the size of the female, and he has much larger and prominent canine teeth along with other distinguishing secondary sex characteristics.

are equally vegetarian, and a well-reconstructed fossil sequence of a now-extinct great ape group showed clearly that they retained their vegetarianism after they moved to the savannahs and up to the moment of their extinction after millions of years of evolutionary changes. These inconsistencies and others lead to some degree of doubt for the interpretations made by prehominid-hunter proponents.

We have no information concerning the nature of the ancestral economies until the time of *Homo erectus* between 700,000 and 200,000 years ago. These hominines indeed were big-game hunters according to the ample fossil evidence. But the fossil evidence does not indicate their social behavior or patterns of living, the composition of the hunting parties, or the relative domesticity of males and females. It would be reasonable to suppose that females usually were not included in hunting, since they were responsible for child care, at least in the earliest stages of infant development. This is a biological concomitant and requires no fossil evidence for verification. Beyond ministering to infants and young children, there is no reason to believe that females never participated in hunting or that they were routinely excluded from such activities. Even in the wolves, which so often are compared with human social groups, nonpregnant and nonlactating females engage in the hunt along with breeding and nonbreeding males. The lioness is the hunter as well as bearing and caring for the cubs of the species, so that mutual exclusion of hunting or other means of obtaining food and of gestating and lactating is not indicated as a required condition for females of all species. The reduction in sexual dimorphism during human evolution argues against the notion that differential sex role assignments are an ancient genetic heritage of the species.

Studies of Subhuman Primates

Relatively few primate species have been observed in the wild, but from these studies we have a varied picture of social and reproductive behavior patterns. Most of the species that have been observed in their natural habitats belong to the group of Old World monkeys, such as baboons, macaques, and langurs. Penetrating reports have been published by Jane van Lawick-Goodall of free-living chimpanzees, and George Schaller has recorded his observations of small groups of the mountain gorilla. Although each species is compartmented into social groups exclusively, group size varies along a broad spectrum. The chacma baboons may live in groups of about 100, while the hamadryas baboon of the dry savannahs may only live in small families that include one male, several females, and young juveniles and infants.

Dominance hierarchies and territoriality are characteristic of the higher primates, but the rigidity of these social constraints is varied for different species. Baboons and macaques exist in more hostile surroundings, and they tend to exhibit a more highly stratified dominance order in which one or more males have *alpha* status and serve to regulate the spectrum of social and sexual activities of the group. The pattern is adaptive for species that live with constant danger, since discipline and order are maintained to the benefit of the whole group. Tree-living monkeys are not so constrained, and each individual can scamper for the nearby trees when a distress signal is given.

Males only approach females in estrus, and the female will accept the male only when she is receptive physiologically. In more rigid societies, such as those of baboons, the alpha males consistently claim prior rights to receptive females and actively prevent the peripheral lower-status males from gaining access to females in estrus. Sexual behavior is much more relaxed in the chimpanzee and gorilla groups, in accord with a more relaxed dominance hierarchy. In his book, Schaller records an incident in which a male stranger entered the group and approached a female in estrus who thereupon presented herself to him. Copulation proceeded in full view of the dominant male of the group, who expressed little apparent interest in the activity. Since gorilla bands usually have only about 10 or fewer adults, a male gorilla may experience no sex for a year or longer. Clearly, sex is not the overriding theme that bonds together the individual members of the gorilla group. Chimpanzees tend to be more active sexually than gorillas, and in these groups there is a similar relaxed dominance arrangement. The dominant male does not exercise strict first choice of females in estrus in either the chimpanzee or the gorilla social groups, in contrast with behavior in baboon species.

The most constant aspect of behavior that keeps individuals together in the group is parental-filial relationships, especially care of the very young. There are variations on this basic theme, as on all others related to behavior. The tree-living langurs entrust the females with the care of the young, whereas male macaques and baboons pay attention to the young and contribute toward their education as future cooperating members of a social group that depends on effective interactions for species survival. There are differences even between species of the same genus; the pig-tailed macaque mother is very stern, while the stump-tailed macaque mother is permissive with her offspring. Intraspecific differences also have been observed in relation to social group structure as well as habitat preferences. Langurs in northern India and Ceylon live in nonhierarchical groups of both sexes and all ages, whereas langur groups in southern India were observed to occur in troops of females and their young plus one adult male, with all-male bands living in separate groups between the female troops and occasionally providing a replacement for the troop male after a skirmish. Other intraspecific differences also have been noted for various groups in different habitats. In general, the social development of the young results from complex sequences of social experience and learning, superimposed on the genetic potential of the species. The development and establishment of adult social and reproductive behaviors reflect numerous inherent and acquired influences during the early formative years of the individual animal in the context of the particular social group in which it has grown up.

Interpretations of Primate Behaviors. It is obvious that we share an ancient heritage with other primates, since the entire order is descended from common ancestral stock. In the same way that 60 million years of evolution led to profound diversification among the primates, we would expect that our own lineage incorporated significant differences from other primate families in the past 15 million years. The pongids and the hominids went along separate evolutionary pathways more recently, which is reflected in numerous biochemical similarities that indicate particular genes shared by the two groups in relatively unchanged forms. In contrast with this, the monkeys diverged very early in evolution from the remainder of the anthropoids. During those 40 million years of separate evolutions, more substantial differentiation has taken place.

Despite the span of time, comparisons continue to be made between human social orders and those of particular baboons and macaques to the point of establishing these monkey societies as models for explaining human social organization and behavior. It is curious that other monkey species with less rigid dominance hierarchies rarely serve for comparisons, nor do the relatively easy-going chimpanzee or gorilla. The frequently cited reason for selecting baboon and macaque models is that these species are savannah residents and thus are more comparable to

Sex Roles in Society 243

the earliest prehominids, who left the trees for the open grassland habitats. The similar ecologies of monkey and prehominid savannah dwellers is implied, hence their similar behaviors in succeeding in this type of living zone. But the prehominid supposedly evolved into a predator, while the monkeys continued to pursue a foraging life-style. Perhaps we should ask whether these differences would influence divergence or continued similarities in social structure. If rigid male dominance was an adaptive feature of social groups on the savannahs, then it is difficult to understand the selection pressures that led to decreasing sexual dimorphism in hominids and a reduction in the defense apparatus of the canine teeth.

In all evolutionary studies and comparisons it is important to distinguish between similarities that arise from a common genetic background (**parallel** evolution) versus superficially similar forms (**convergent** evolution) with entirely different genetically specified programs that underwrite the characteristics being studied (Fig. 11.7). We have no knowledge of the genetic comparability of social organizations in different primate

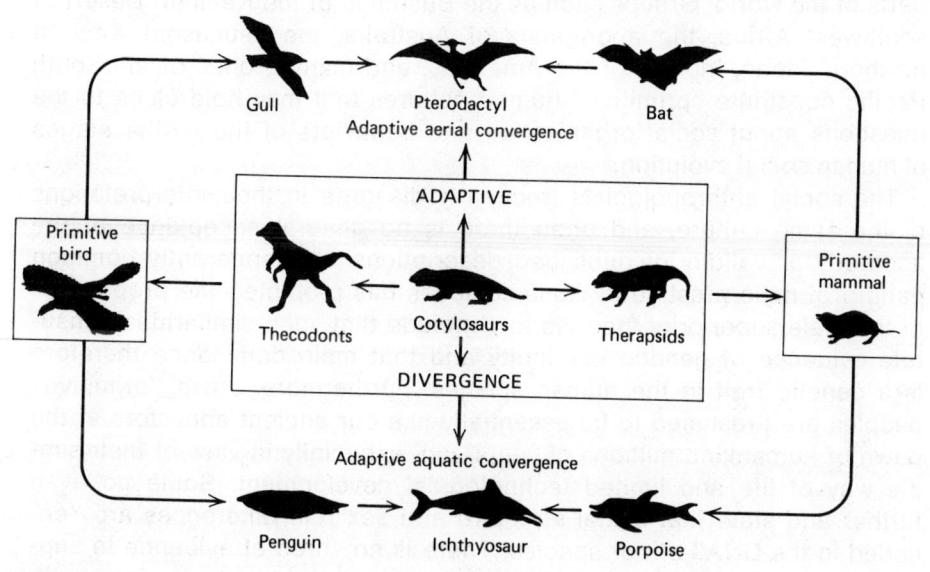

figure 11.7
The evolutionary phenomena of adaptive modifications. *Divergence* leads to varied forms from a common ancestor, and continued evolutionary changes may lead to a number of distantly related species that have similar life styles. Such *convergence* produces superficially similar forms, but examination of the development of structure and function generally reveals their substantial genetic differentiation. In *parallel evolution*, the similarities result from modifications in the same or similar genes, whereas convergence results from modifications in different genes that regulate similar structures and habits.

species. But it is obvious at the same time that sociality is a trait that the higher primates have in common and thus must be specified in the basic genetic program. It is difficult to determine the particular selection pressures that contributed to social and sexual behavior modifications, and superficial comparisons between selected species for selected traits merely obscure the search for the origins of behavioral modes in whole families of species. A more honest comparison should include intraspecific and interspecific variations, as we know to exist for those primate species that have been studied, such as macaques, baboons, langurs, chimpanzees, and others. If one selects particular items of information and ignores the rest, a coherent picture may emerge, but this coherence is artifactual and may itself be a distortion of the truth.

Cross-Cultural Human Social Patterns

In addition to viewing modern cultures on all the continents and noting the overwhelming exclusion of women from prestige activities, many comparative studies have been made of preliterate peoples in various parts of the world. Groups such as the Bushmen of the Kalahari Desert of southwest Africa, the aboriginals of Australia, the caucasoid Ainu of northern Japan, Indians of the Americas, and many groups of the South Pacific constitute "primitive" human cultures that may hold clues to the questions about social organization and behaviors of the earlier stages of human social evolution.

The social anthropologists frequently disagree in their interpretations of the same culture, and often there is no general acceptance of the accuracy or validity of published descriptions. The apparently common pattern of male prestige in these societies has prompted the proponents of the male superiority theories to conclude that such similarities constitute evidence of genetic continuity and that male dominance therefore is a genetic trait in the human species. Furthermore, these "primitive" peoples are presumed to be essentially like our ancient ancestors at the dawn of humankind millions of years ago, especially in view of their simple way of life and limited technological development. Some go even further and state that social structure and sex role differences are "encoded in the DNA" of our species. There is no shred of evidence to support these glib conclusions, nor are they acceptable on any genetic grounds.

If there is a genetic continuity to human social order then one must be prepared to explain the astonishing speed with which old patterns have been abolished and new ones established firmly in their place in the span of a single generation. Such rapidity of change for complex genetic traits is unknown, and impossible from a theoretical genetic viewpoint. The facile citations of DNA and natural selection in support of these absurd notions only underscore the specious nature of the reasoning that has been employed. In addition to our singular lack of information concerning the genetics of human behaviors, male supremacists compound their casuistry by citing "evidence" derived from an arbitrary sampling of particular societies that fit their preconceived notions. The usual "models," which are claimed to comprise the evidence for genetic continuity of male dominance in human societies, frequently turn out to be those that just happen to demonstrate extremes of difference between male and female roles. There is no question that these selected cultures include virile males and timidly domestic females, but these hardly can be construed as the norm or model of ancient humanity. The hundreds of preliterate cultures that have been studied demonstrate a tremendous spectrum of variability in individual social and sexual behaviors. The selection of a norm or model cannot be anything but arbitrary and self-serving in purpose.

Behavioral Plasticity

Human behavior is unique among animals in lacking an inherent set of responses to the same stimuli. No matter which group of baboons we may observe, a subordinate male does not stare full in the eye at the alpha male. In human societies the confrontation varies, so that looking squarely at another person may be deemed extremely impolite in some groups and shifty in others; it led to death for an Aztec who gazed on the emperor, but it is considered a democratic prerogative for other cultures. Human beings gradually acquire the habits, skills, and beliefs that integrate them into society by processes of socialization and acculturation. Except for a few specific biochemical traits such as blood group factors, most genetic information is modulated by environmental interactions to produce a varying spectrum of expressions. The greater the variability in expression the more influential the environmental modulations of the genetic program of potentials. Even our fingerprints are different for each of the ten digits, although a single genetic program is present in an individual's DNA.

The incredible plasticity of human cultural behaviors indicates that most of these traits are learned in response to societal conditioning. An infant can become a member of any society after learning the language and customs during its period of growth and development. Depending on the interactions between genes and environmental experiences, one individual may be as intelligent, cooperative, or strong as any other with a similar genetic potential. A male growing up in New Guinea may become subordinate to the female if he is in the Tchambuli tribe; he may be equally aggressive as the female in the Mundugumor tribe, or equally nonagressive as the female if he acculturates into the Arapesh tribe.

There is some expression of gender dimorphism in virtually every human society that has been observed, but these expressions vary through a spectrum from the extreme of male dominance to the other extreme of female dominance. There is no evidence for the presence of a fixed "biogrammar," as proposed by Lionel Tiger and Robin Fox, by which our behaviors are directed according to coded instructions in DNA. Instead of considering the notion as hypothetical, these authors have elevated their idea to the level of fact. The evidence that is available points instead to the particular development of societal sex roles under the influences of conditioning to some set of expectations and the *acceptance* of these expectations, superimposed on the biological verities of different reproductive capacities for males and females. Women conceive, bear, and nurse children, which thus determine a substantial component of female participation and contribution to preliterate and literate societies alike. The development of gender identity as masculine or feminine centers around this biological theme in many cultures. We shall now consider some of the lines of evidence which bear on the relationships between biological sex and the acquisition of male or female identification in a social context.

Development of Gender Dimorphism

Numerous observations and experiments have been made using a variety of vertebrate animal species. Humans have been the subject of many psychological, sociological, anthropological, and clinical studies conducted in efforts to trace the sequences and consequences of individual identification with one sex or the other and with the expression of such **gender identity** in relation to fulfilling a particular role in one's life and in society. The interpretations often are controversial, usually based on incomplete or inadequate evidence, but the overall impression that results is that of a combination of biological and cultural inputs to the final product of an unambiguous gender identity for an individual person. The interactions of biological and cultural factors direct the development of gender identity in an individual and lead to its expression as more masculine or more feminine in the male-female continuum of behaviors. The specific set of cultural traits by which each gender is recognized differs among human societies, but each society has an established pattern for the maintenance of some system of gender differentiation.

Development of Biological Sex

Although we discussed this topic earlier in Chapter five, it can be summarized briefly here in the context of sex versus sex role assignments.

Each human embryo contains the genetic blueprints for development as a male or a female, but this development does not begin until the sixth week of embryonic life. At this time the embryo that is **XY** begins to differentiate two testes from the gonad rudiments, which are identical in all embryos up to this time. If the embryo is **XX** then gonad differentiation as ovaries will be delayed until about the twelfth week. The first differentiation of sex thus involves the specification of the gonads as testes or ovaries, according to the sex chromosomes present in the embryo.

Once the testes or ovaries have begun to differentiate, hormonal secretions are produced, and these in turn influence the development of one set of internal ducts. Androgens produced by the testes influence the development of internal reproductive structures typical of the male, whereas estrogens in the female embryo direct female duct differentiation. If the gonads fail to develop, or if the embryonic cells are insensitive to androgens, then female internal structures will differentiate exclusively. In other words, female differentiation will take place in the presence or absence of ovaries, and in **XY** as well as **XX** embryos if androgens are deficient or ineffective. Androgens are required to direct male differentiation, but these hormones must be present at particular and crucial times during embryonic development.

The final step in the development of basic sexual anatomy involves the differentiation of the external genitalia during the third month of fetal life. This structural differentiation depends on the earlier embryonic events of gonad specification and on the hormonal repertory in the fetus. From clinical and experimental evidence it appears that differentiation of female reproductive and external genital structures occurs independently of the presence of ovaries and ovarian hormones. Male internal and external reproductive anatomy is androgen-dependent, the male sex hormones normally being produced in the highest amounts in the embryonic testes.

Clinical Studies

There are two particular inherited conditions that lead to the development of external genitalia that are not matched to the sex chromosome and gonad constitution of an individual. Patients who are **XY** and possess functional testes may develop female external genitalia if their cells are insensitive to the androgens that they produce in normal amounts. The body becomes feminized in appearance because of the estrogens, which also are produced in normal testes, and which are not masked by androgen action. The other inherited condition occurs in individuals who are **XX** and have ovaries and female internal anatomy, but who become externally masculinized to varying degrees because of androgens produced by the adrenal gland, instead of the usual cortisone hormone. A similar condition may develop if an **XX** fetus is exposed to androgens in the maternal bloodstream, caused either by an androgen-producing tumor or by administered hormones that mimic the action of androgens on developing fetal organ systems.

John Money and his colleagues have reported a number of case and group studies of masculinized **XX** and feminized **XY** individuals of different ages, and while there is less scope of comparison groups and age distribution than one would like, some interesting findings have been described. Among 48 individuals who were fetally androgenized, there was clear evidence of "tomboy" behavior during childhood along with other traits (mode of dress, toy preferences, etc.) that form part of this stereotype. These individuals were reared as females following hormonal and surgical therapy either in very early childhood or later in adolescence and adulthood. The majority showed no evidence of masculine gender identity or role, even when high androgen levels were continued into adulthood and prior to therapy. Postnatal influences predominated in the determination of final gender identity, even though some prenatally influenced behavior was detectable as "tomboyism." A group of ten androgen-insensitive **XY** individuals were studied, all have testes, but female external genitalia; the external morphology led to their assignment as females at birth and to their subsequent rearing as females. The testicular estrogens promoted feminization of the bony structure and outer contours of the body, including typically female growth of the breasts, but since neither the male nor the female internal ducts had differentiated, these individuals were sterile. All ten who were studied displayed stereotypical feminine gender identity and an anticipated and desired female role expression. Except for the fact that they could not bear their own children, these ten individuals were female in all behavioral traits, even though each was of the **XY** karyotype and possessed testes instead of ovaries. A particular gender identity and role clearly can become established according to postnatal determinants, even when chromosomal, gonadal, and hormonal traits may be of the opposite sex.

One case study described by Money was particularly interesting in that it involved identical twins who were assigned different sexes and reared as a boy and a girl. One of the boys suffered a traumatic accident at the age of seven months that led to the loss of his penis. Upon medical advice, the parents agreed to sex reassignment for the child and to the surgery and hormonal therapy this entailed. The children were exposed to very different treatment by their parents; it was geared to learning to behave according to the assigned genders. The boy behaved as we have come to expect, that is, active in play, less restricted in tidiness demands, mischief more tolerated by his parents, and so on, and the girl was encouraged to engage in domestic activities with her mother, to tone down the rough-and-tumble play activities, and to adhere more strictly to "good manners," among other traits. At the time the last observations were made when the twins were six years of age, they had clearly established gender identities and expectations of a boy and a girl. Since these children were genetically identical, it is obvious that sex assignment and rearing exerted significant and overriding influences on the development of masculine and feminine gender identities. The notion of a "biogrammar," which directs a different sex role development in males and females according to coded instructions in DNA, seems unlikely in view of this particular case history. One isolated example does not prove a point, but it does lead us to question the rigidity of male supremacist theories.

From studies such as these, and the countless articles that have been published in books and journals of psychology and anthropology, it would seem that each individual of a social group is conditioned to expect and to accept a very different gender role for males and females. Cultural conditioning begins in infancy and is reinforced continually by different behaviors directed toward the individual by other children and adults of the society. This acculturation provides the particular framework by which an individual gains a masculine or feminine gender identity and then expresses the identity in social and sexual behaviors, which are different for males and females. The existence of differences does not imply the superiority of one group relative to another, unless we learn and then accept such an interpretation. These acquired beliefs underlie the myths of sexism, racism, ethnicism, and other credos of inherent superiority of one kind of person relative to another. In a relatively brief time we have disavowed the superiority of landowners to the landless, of the nobility to the commoners, the employer to the employee, and numerous other paired situations. It remains for us now to discard those remaining insidious patterns of thought and behavior that cheat the individual of dignity and that prevent a substantial proportion of our population from making maximal contributions to society and each other.

Suggested Readings

Alland, A., *Evolution and Human Behavior*. Garden City, N.Y.: Natural History Press, 1967.

Alland, A., *The Human Imperative*. New York: Columbia University Press, 1972.

Avers, C. J., *Evolution*. New York: Harper & Row, 1974.

Fox, J. R., The evolution of human sexual behavior. In *What a Piece of Work is Man* (eds., J. D. Ray and G. E. Nelson). Boston: Little, Brown, 1971, pp. 273–285.

Freedman, D. G., A biological view of man's social behavior In *Social Behavior from Fish to Man* (by W. Etkin). Chicago: University of Chicago Press, 1964, pp. 152–188.

Howells, W. W., Homo erectus. *Scientific American, 215* (November 1966), 46.

Jost, A., A new look at the mechanisms controlling sex differentiation in mammals. *Johns Hopkins Medical Journal, 130* (1972), 38.

Lipman-Blumen, J., How ideology shapes women's lives. *Scientific American, 226* (January 1972), 34.

Mead, M., *Male and Female.* New York: Morrow, 1949.

Money, J., and A. A. Ehrhardt, *Man & Woman, Boy & Girl.* Baltimore, Maryland: The Johns Hopkins University Press, 1972.

Montagu, A., *Sex, Man, and Society.* New York: Tower Publications, 1969.

Morris, D., *The Naked Ape.* New York: McGraw Hill Book Company, 1967.

Napier, J., The antiquity of human walking. *Scientific American, 216* (April 1967), 56.

Pfeiffer, J. E., *The Emergence of Man.* New York: Harper & Row, 1969.

Simons, E. L., The early relatives of man. *Scientific American, 211* (July 1964), 50.

Tiger, L., and J. R. Fox, *The Animal Imperative.* New York: Holt, Rinehart, Winston, 1971.

Weckler, J. E., Neanderthal man. *Scientific American, 197* (December 1957), 89.

chapter twelve

Population Dynamics

Along with other large primates, the potential for increase in the human species is limited by several biological factors: (1) usually one infant per birth, (2) approximately one-year intervals between pregnancies, and (3) period of fertility during the life span. Under favorable conditions, the human population could double every 17 years. Yet this theoretical rate of increase actually has not been realized either in our past history or at the present time. There clearly must be checks on population growth that influence the particular rates at any time or place and also regulate the rate at which rate-increases themselves will fluctuate. Two of the important elements of population dynamics that must be inserted into a proper understanding of flux are the *birth rate* and the *death rate*.

Birth Rates

There are three main factors that influence the rate of newborns produced: **age distribution of women** in the population, **marriage patterns**, and **birth control programs**. The interval of fertility in women generally occurs between the ages of 15 and 50, with the highest fertility rate in the midtwenties. The higher the proportion of women in the years of child-bearing age, the higher the birth rate, with an additional factor of increase if most of these women are in the earlier age groups (Fig. 12.1). In newly settled areas there may be an initially high rate of births if the majority of people are younger, whereas regions comprised largely of older people would obviously exhibit low birth rates. Patterns of emigration and immigration thus would be correlated with differences in birth rates, as one obvious example of the operation of this factor.

Trends in time of marriage and in the proportion of individuals who marry and establish families have been related specifically with population flux. In France between 1750 and 1850 it has been estimated that 30 to 40 percent of women remained unmarried and 60 percent of males did not establish families. There was a "birth crisis" in eighteenth-century France that received considerable attention, and it was largely because of the low frequencies of marriages and births. A similar pattern was characteristic of much of western Europe at that time. The average age at which women were married has been estimated as under 20 in Europe during the fifteenth century, but in the mid- to late twenties during the eighteenth century. The marriage age generally has been correlated with economic factors, later and fewer marriages occurring in times of economic depression than at other times.

Birth control probably has been practiced ever since the human mind first discovered the relationship between coitus and conception, with coitus interruptus (withdrawal) still considered to be the primary means for regulating family size in most of the world. Three particular lines of evidence provide support for the influence of contraception on declining birth rates. In countries such as Holland where comparisons can be made between people of different religions, it has been documented that Catholics produce larger families than Protestants. Where vigorous campaigns for birth control have been instituted, there generally is a corresponding decline in birth rate. The introduction of new contraceptives, such as the pill or intrauterine devices, also has been coincident with a significant decrease in birth rate where the means have become available to a sufficient proportion of the population.

A survey of birth rates in different parts of the world reveals that near maximum rates prevail for much of the world's population, although there is a declining rate in some countries and a rising rate in others. Most of the underdeveloped nations demonstrate high birth rates, which are associated with early marriages and no programs for limitation of family size, as in India. In Japan, on the other hand, an effective birth control program leads to limitation of family size and a generally lower birth rate. Approximately one in three pregnancies is terminated by abortion, which is free and available for all to utilize under Japanese government auspices.

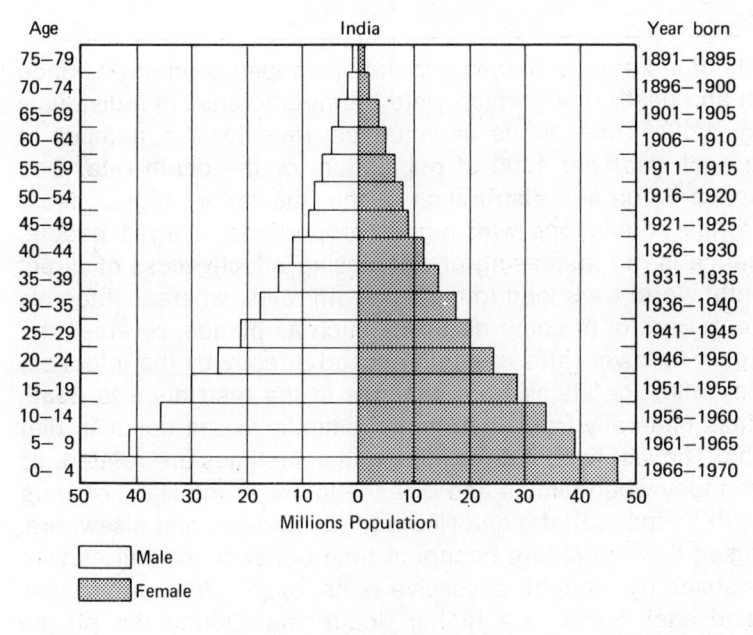

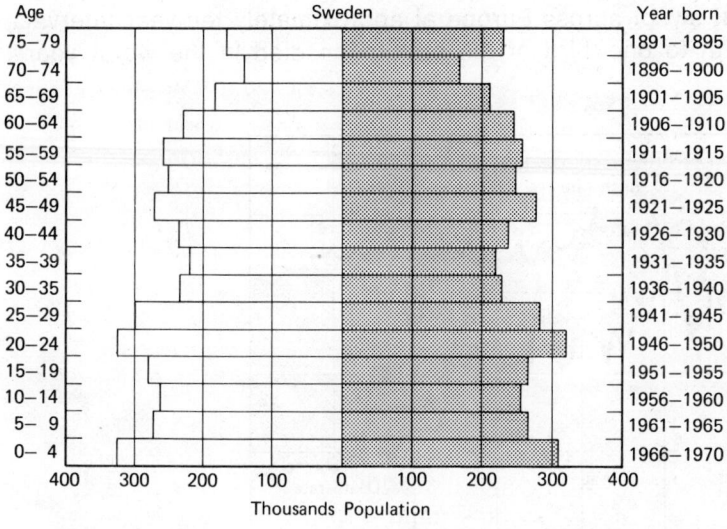

figure 12.1

By charting populations according to sex and age it is possible to make predictions about the future growth rate. If the birthrate were to be reduced to one-half of its present level in India, there would be zero population growth by the year 2040, but this would include well over one billion people by the time the size leveled off. In Sweden, on the other hand, no population change will occur unless there is a dramatic increase in birthrate from its present level.

Death Rates

The actual rate of increase in human populations depends on the balance between birth and death rates, which yield the net increase in individuals per year (Fig. 12.2). Three kinds of influences monitor the number of people dying each year per 1000 of population, or the **death rate**. One obvious influence is the **age distribution** of the population; higher death rates characterize populations with higher proportions of aged people. Another influence is the increasing or decreasing **effectiveness of direct causes of death**; world wars lead to higher death rates, whereas intervals of peace do not. Control of some diseases, such as plague, reverses the trend from higher to lower rates of death caused directly by the infection. A third kind of influence stems from **changes in the resistance to death** in the individual, generally from such improvements as can occur in diet or in sanitation. The last two kinds of influence sometimes are related, as we know even today when famine and disease follow in the wake of wars such as those in Vietnam, Biafra and Nigeria, Bangladesh, and elsewhere.

Famine caused by crop failure occurs in time of peace as well as war, and may be caused by drought, excessive rains, or infestations by pests or disease, and each leads to a higher death rate. During the plague epidemics that swept across Europe at approximately ten-year intervals, from one-fourth to one-third of the population died in the worst years.

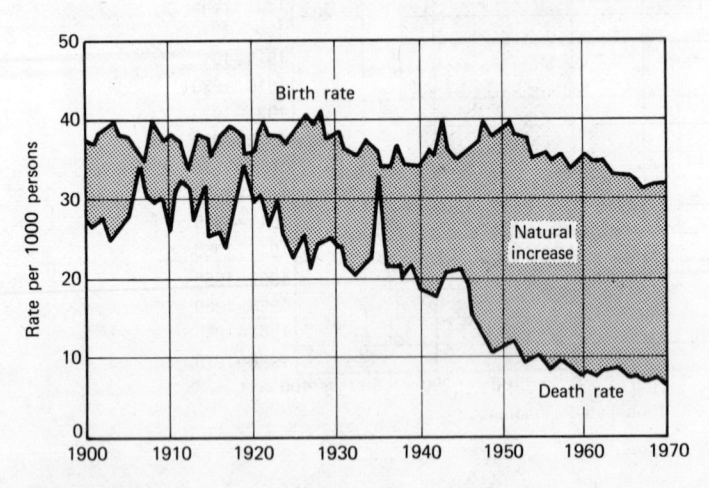

figure 12.2
The rate of natural increase in population depends on the relative difference between rates of birth and death. In this chart taken from Ceylon statistics, it is clear that overall fluctuations have occurred in a 70-year period but that population increase is greater now because of a declining death rate even though relatively little change in birthrate has taken place.

During the latter part of the eighteenth century, smallpox accounted for 10 percent of the deaths in western Europe. The introduction of a cowpox-containing vaccination method by Edward Jenner in 1798 was directly responsible for the reduced infant mortality rate from smallpox infections, and thus for a general reduction in the population death rate by the beginning of the nineteenth century. Indeed, since 1800 there has been a worldwide decline in the death rate, principally because of improved health services and second by virtue of improved diet.

The basis for comparing mortality rates is taken from calculations of the **expectation of life at birth.** Evidence from mummies indicates that upperclass Egyptians had a life expectancy of 25 to 30 years, about 2000 years ago. Europeans during the Middle Ages lived an *average* of 35 years. The significant feature of life expectancy values lies not in the average, but in the death rates at different ages. Where there is a high infant mortality rate, there is a higher average death rate for a population, but for those who survive infancy there is a greater chance of living up to the time of the next peak in death rate (Fig. 12.3). Lowered mortality rates generally occur by lowered rates of death in infancy; increasing *longevity* is another situation altogether, but infant mortality and longevity are the primariy components used to calculate overall mortality rates in populations.

In the developing countries there is a lower life expectancy (as well as a higher infant mortality rate) than in the developed nations of the world. The average life expectancy in India in 1931 was 25, whereas for a person living in England, it was 59 years of age. There has been a steady increase in average life expectancy, but the *rate* of increase is higher now than it was a century or more ago. Some available figures for men in England show an average life expectancy of 40 years in 1841, 44 in 1891, 59 in 1931, and 78 in 1971. The rate of increase was 10 percent between 1841 and 1891, but rose to 34 percent between 1891 and 1931, and remained at about this rate during the next 40 years. If this same approximate rate continued, then an Englishman's average life expectancy would be about 104 by the year 2011.

Population Growth

The pattern of world population growth has changed at various times during the past 25,000 years (Fig. 12.4). There were an estimated 3 million people on the Earth 25,000 years ago, and this had increased to only 5 million 10,000 years ago. There had not been even a doubling in the intervening 15,000 years. But we believe that agriculture was introduced into many human societies about 10,000 years ago and that profound

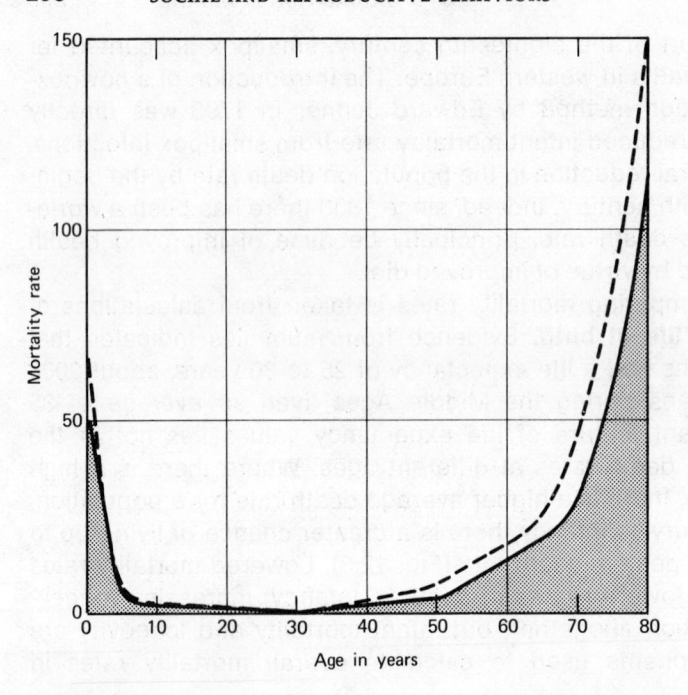

figure 12.3
The mortality rate varies according to age and is somewhat higher for males than for females at particular times in life. This graph is typical of countries such as the United States and Great Britain at the present time. The female mortality rate is shown as a shaded area, and the dashed line represents the mortality rate for males.

changes subsequently became possible. Smaller areas could sustain larger populations than was possible for hunters and food-gatherers, and settlements arose where life previously had been primarily nomadic. Urbanization was initiated by this agricultural revolution, and adequate food supplies were provided by cultivation for these growing numbers of people. The estimated population of the world increased from 5 million to 86 million in the next 4000 years, and to 133 million in the 4000 years after that. During the 8000 years after the beginnings of agriculture, the population underwent a 25-fold increase, but the greater rate of increase occurred in the first 4000 years of this period. Urbanization developed apace with population increase, and communities were characterized by a greater division of labor than could be achieved by hunters or other nomadic peoples. One person produced more food than was required for his own family's needs, thus freeing some people to pursue other activities and still secure an adequate supply of food by purchase or barter.

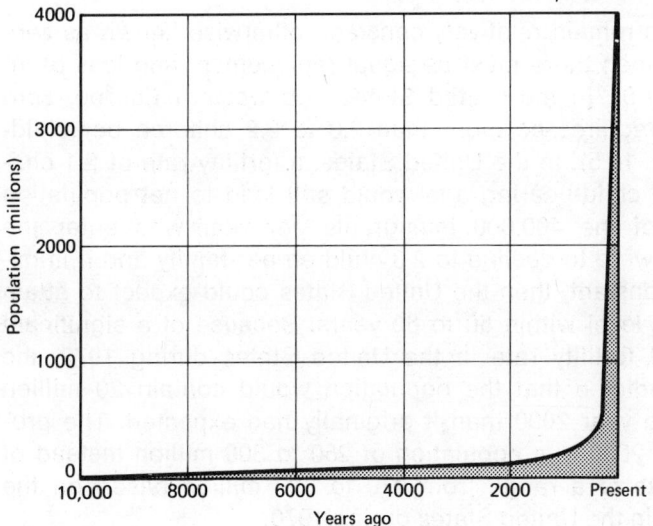

figure 12.4
The changing rate of population growth in the world during the past 10,000 years.

By 1650 there were about 500 million people in the world, and it had taken about 1000 years for the population to double to this size. The next interval of doubling of world population only took 200 years, and there were a billion people in the year 1850. The next doubling took only 80 years, so that 2 billion people were in existence by 1930. In 1950 the figure rose to 2.4 billion; there were 3.8 billion in 1972; and there will be 6270 million people on the Earth by the year 2000 if the present rate of increase is maintained. The rate of world population growth is now 2 percent per year, which is an increase in humans of 1,400,000 per week; 199,104 per day; 8296 per hour; or 138 net increase every minute. These figures lead to the prediction of 4 billion people in 1975, 5 billion in 1986, 6 billion in 1995, and 7 billion in 2025. This is a doubling rate of 37 years for the world, and it is even more staggering to realize that it was only in 1850 that we reached the 1 billion mark!

The rate of doubling varies from one country to another, so that it is 20 to 35 years for developing countries but 50 to 200 years for the more developed nations of the world. Some population doubling times are: Mexico, 20 years; Brazil, 22 years; Pakistan, 23 years; Egypt, 24 years; India, 28 years; Nigeria, 28 years; Japan, 63 years; United States, 70 years; Italy, 117 years; and Great Britain, 140 years. These statistics assume maintenance of present-day rates of population increase. If the population size is to remain relatively constant, otherwise known as *zero population growth*, then there must be equal replacement and loss of individuals (ratio of 1.0). In the United States and western Europe, zero population growth requires no more than 2.0 to 2.2 children per childbearing couple (Fig. 12.5). In the United States, a fertility rate of 2.1 children per woman of child-bearing age would still lead to net population increase because of the 400,000 immigrants per year who enter the country. If this rate were to decline to 2.0 children per family and if immigration remained constant, then the United States could expect to attain a stable population level within 50 to 60 years. Because of a significant decline in the total fertility rate in the United States during 1972, the Census Bureau predicted that the population would contain 20 million fewer people by the year 2000 than it originally had expected. The projection for the year 2000 is a population of 250 to 300 million instead of its previous estimate of a range from 270 to 322 million based on the higher fertility rates in the United States during 1970.

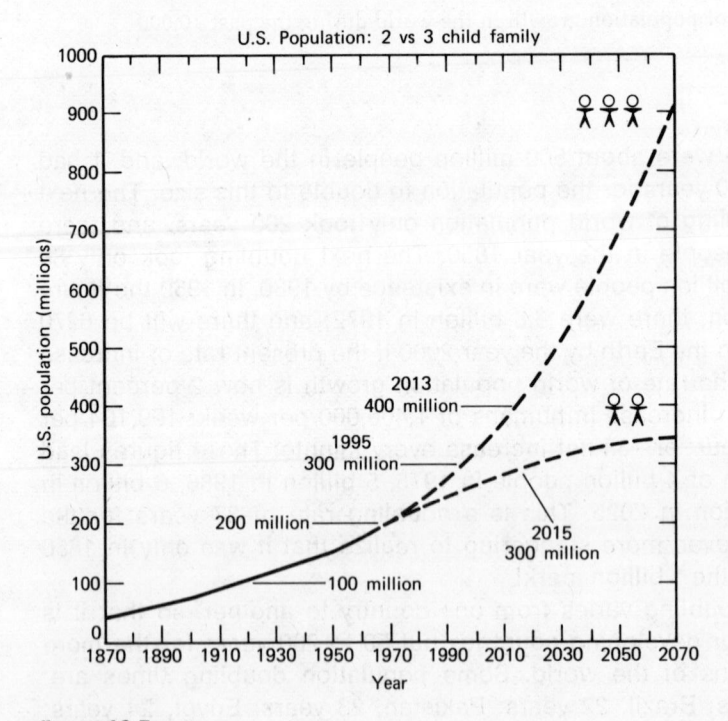

figure 12.5
Different predictions for population growth in the United States, depending on the birthrate. Zero population growth could result if there was an average two-child family, but this would not stabilize until about 50–60 years from now.

Factors Affecting Population Growth

During the 100 years between 1750 and 1850 the population of Europe approximately doubled. Midway in this period Thomas Malthus wrote his classical essay on population growth, citing various factors that regulated population increase. Chief among these were the primary scourges of war, famine, and disease. He also listed subsidiary factors, but did not discuss these in as great detail. A recent article by historian William Langer, published in *Scientific American* magazine, provided some interesting insights into events that contributed to the population growth pattern in Europe during that century. It may be helpful in this general discussion to examine some of the points made in that article.

Primary Regulating Factors

Although wars were fought during the eighteenth century in Europe, there was a substantially lower toll of the population in the 100 years that intervened between the Napoleonic Wars and World War I, so that at least part of the population increase between 1750 and 1850 probably was due to a slackening of this means of population reduction. With a lesser effect of wars, there also was a lesser effect of the famine and disease that accompany the conflicts between armies, but that principally affect the general population. Famines because of crop failure were less frequent between 1750 and 1850. With the development of better means for communication and transportation over greater distances, food shortages could be alleviated more easily than in earlier times in Europe. According to Thomas Malthus, populations tend always to rise to the level of available subsistence, but while food increases *linearly* (1, 2, 3, 4, etc.) population doubling is *exponential* (2, 4, 8, 16, etc.). The increase from 140 million in 1750 to 265 million Europeans in 1850 probably also was feasible because of new food sources, as well as improved distribution channels. It was about the middle of the eighteenth century that the potato began to be cultivated extensively in Europe after its introduction there from America; it is raised easily even in poor soil, and small plots are sufficient. The potato is the main food supply to sustain families even today. Langer suggests that the potato may have been one of the most influential factors leading to the dramatic increase in population after 1750. We know that the potato famines in nineteenth-century Ireland prompted wholesale emigrations, thus further indicating the overriding importance of this food resource in population flux.

Death from disease continued to be an important factor in keeping down the population size, but with decreasing incidence of plague and smallpox, these two principal killers accounted for a smaller percentage of mortality and thus a general decline in deaths due directly to disease in the nineteenth century. Improved medical services, and sanitation to a lesser degree, also contributed to reduced mortality and thus to population increase in Europe.

Subsidiary Regulating Factors

Two additional factors that influenced European population growth were *marriage customs* and *infanticide*. Celibacy was very common in western Europe before 1850, enforced by legal and economic restrictions of various kinds. Poor Laws were designed to prevent paupers from propagating more paupers, permission to marry was given infrequently to the lower social classes by their employers, and economics affected the size of families that did develop.

Infanticides still accounted for about 6 percent of all violent deaths in England as late as 1878, and were more frequent earlier. The abandonment of infants was so common that government subsidies eventually were provided for foundling homes and later withdrawn as the budgets became strained in some regions. The first foundling hospital in London was opened in 1741, government funding permitted open admissions by 1756, and funding was withdrawn in 1760 because the costs were so high. Vincent de Paul helped to establish the first foundling hospital in France in the latter part of the seventeenth century, and such institutions were established elsewhere on the continent during the eighteenth century. From 1824 to 1833, the records show that 336,297 infants were left at French hospices for foundlings. Since an infant death rate of 90 percent was known for these foundling hospitals, abandonment was essentially equivalent to the direct methods commonly used by parents to kill unwanted infants.

Although marriage customs and infanticides accounted for some checks on population growth before 1850, neither could stem the tide of increase that characterized that era. The high rate of population increase led to widespread fears of overpopulation in Europe, but these concerns ultimately were relieved by the introduction of industrialization on a large scale and by the emigration of millions of Europeans to other parts of the world.

Some Current Concerns of Overpopulation

The decrease in infant mortality during the past 300 years led from a 50 percent chance against infant survival to the present average in the developed countries of 2 percent chance that the infant will not live to be one year old. The infant mortality rate in the United States is higher than for many other modern nations, primarily because of inadequate medical services for the urban poor. We rank fifteenth in the control of infant mortality in the world. Still, the rate is relatively low and together with increasing longevity, global population increase would continue to escalate unless checks were instituted. What are some of the consequences of increasing population, and what can be done to reduce this increase or to stabilize the world to the level of zero population growth? We will consider these questions briefly.

Some Consequences of Population Increase

About 2 billion out of the 3.8 billion people in the world today are undernourished. The one third of the nation that were "ill-clothed, ill-housed, and ill-fed" in the depression days of Franklin D. Roosevelt's administrations still exist, and congressional committees have estimated that 20 million citizens of the United States live at the starvation level. Despite the high productivity of American farming, about 20 million acres of fertile land are not planted. The government rents these acres and pays the farmers not to grow food on the available land. Similar programs exist in other countries, such as Argentina and Canada. The economics of the world market are not premised on the statistics of undernourishment or outright starvation. Even though an adequate food supply could be grown theoretically, world policies mitigate against a proportionate distribution of these supplies around the globe. Countries that can buy wheat are able to supplement their food needs; countries too poor to buy the bounties in the storage silos remain essentially unaided by the "have" nations. Improved agriculture in India will aid the Indian food supplies, but improved agriculture in the United States or Canada has no direct effect on India's problems. The prognosis for such "have-not" nations remains poor, and their people will continue to live in relative deprivation if the population increases and the means of subsistence remains low. The dilemma becomes more aggravated as exponentially increasing populations diverge further away from the linearly increasing food supplies.

The Earth contains a finite set of resources, and affluent nations consume a disproportionately high percentage of these resources relative to their population size. The United States consumes 35 percent of the world supplies of fossil fuels (oil, coal, and natural gas), but has only 6 percent of the world's population. Minerals, timber, and other natural resources are being depleted rapidly by the highly industrialized nations to satisfy an affluent consumer demand, at the expense of countries that are undeveloped or underdeveloped. Conflicts for these dwindling resources become more probable, and the prospect of wars between technologically advanced societies conjures up images of wholesale slaughter and destruction.

The continuing economic growth of the developed countries proceeds with little or no management at the national or international levels. The goal of ever-increasing affluence for the affluent has led to global problems of pollution, declining irreplaceable natural resources, and regulation of the quality of our life by huge industrial corporates essentially unhampered by legal restrictions. The legislative controls that were introduced to prevent the abuse of labor by management had ameliorating effects, although we may have swung toward an opposite extreme today as we view labor's manipulations of the general population by an unending series of strikes and work stoppages. As participants in a planetary ecology and economy, we can only hope that international regulations will be formulated so that continued exploitation of the many by a relative few will be outlawed. Growth cannot continue indefinitely if the quality of life is to improve for the world's billions, because resources and technologies are limited. If the developing nations are to be permitted to develop, then the industrialized nations must reduce their pace of increasing affluence and achieve a status quo. We still would maintain a high standard of living, but we would not be so sharply differentiated from other peoples if they are allowed to "catch up." To maintain our goal of an ever-expanding economy and ever-increasing affluence would concomitantly lead to reduced chances for improvement in developing populations. In the future competitions for finite resources, only some of the competitors would survive. For us all to survive and for future generations to inherit a legacy of quality in life, it seems essential that there be international monitoring of the exploitation of the Earth's resources and potentials. If not, then the "have-nots" will increase and only some of the "haves" will remain to bear witness to human folly.

Zero Population Growth

The principal hope for achieving a stable world population size resides in the prospects for effective programs for birth control. There are different kinds of barriers to achieving significant levels of fertility control, but mainly these include ignorance, cost, and the psychology of human actions.

Large families are more common in societies in which there is a high rate of infant and child mortality and a deep concern about security in old age. As these reasons decline, so does the birth rate, especially as the costs of living and the standards of living increase. Penalties may be exacted in countries with the national goal of populational change or stability; these penalties may be higher taxes on unmarried people if the country is underpopulated, or higher taxes for larger families in overpopulated nations, with an additional cost for each child in excess of some particular norm value. Nations that embark on a vigorous program of birth control may ease the availability of medical facilities. Abortions are legal, free, and readily available in Japan, where the rate of population increase was reduced from 30.2 per thousand in 1947 to 17.6 per thousand in 1959, and this lowered rate has been maintained to the present day. The government of India is making a marked effort to provide education and medical facilities that will encourage men to undergo vasectomies, and some success has been achieved in the past few years. There is no governmental action in the United States to back up its presumed desire of stabilizing the population size. The single person is taxed disproportionately, there is a tax bonus for each additional child in the family, educational programs are minimal and poorly financed, and medical services fluctuate according to state laws regarding abortion. These discrepancies between observed and presumed goal-oriented actions were well reflected by individual responses by Cornell University faculty and students to a recent questionnaire concerning family planning in relation to population control. Although 84 percent of the 1059 respondents favored limitation of family size, 65 percent said it wanted three or more children; only 6 percent of the males favored voluntary vasectomy once full family size had been achieved, and a similarly small percentage of females preferred voluntary sterilization to other modes of fertility control after the completion of the planned family. Apparently we prefer the "others" to regulate their lives so that global population will stabilize, but we do not extend the restrictions so eagerly to ourselves.

The simplest, safest, and least costly means for fertility control is male vasectomy. Oral contraceptives are expensive, often unavailable where birth control is desirable, and they alter the entire reproductive physiology of the female who subscribes to this regime. Withdrawal and the rhythm method of fertility control are essentially unreliable, but free to the family and produce no lasting physiological distress. If the whole world depended on these two methods, zero population growth could never be achieved. Although vasectomy would appear to be the most effective of the alternatives, there is very high resistance to its utilization. Most people do not know that the simple surgical procedure of tying off the vas deferens merely blocks the passage of sperm to the ejaculatory duct and has no effect on virility, sex drive, or sexual gratification. The hormonal system remains intact, and erection and ejaculation still precede orgasm, but there are no sperm in the seminal fluids (see Fig. 5.1). In addition, the present low rate of successfully reversing the vasectomy by reuniting the tied ends of the duct mitigates against more frequent utilization of this means for fertility control. Tying off the Fallopian tubes in women similarly prevents later unplanned pregnancies, although a few days of hospitalization are required following the simple surgery.

If zero population growth is to be realized in the world, education and improved medical services at low cost are absolutely essential. But even when these goals are announced and under way, the whims and wishes of the human mind still serve to direct the realization or failure of effective programs for fertility control.

Some Prospects for the Future

One of the voiced concerns arising from the disproportionate rates of reproduction among different socioeconomic classes in our own country or in the world as a whole is whether or not there is or will be a decline in human intelligence. The obvious premise that underlies this concern is the belief that reported differences in IQ test performance will be enhanced as the less advantaged groups increase their frequency in the population by virtue of demonstrably larger average family size. Whether the contrasting groups are of different racial, ethnic, or social backgrounds, there is a common pattern of differences associated with both IQ performance and average family size. There are no unequivocal data on which to base firm conclusions, except to note that the relative inputs from genetic and environmental influences are difficult to separate. Since most IQ tests are culture oriented, they are subject to wide variations in results from one group to another and from one time to another in repeat tests of the same general groups. Such variations obscure the genetic contribution to the results, but they show at the same time that environmental factors are extremely influential in producing particular test result patterns. Unlike a trait such as blood type, which is distinct, unchanging during the lifetime of the individual, and expressed in every case where particular alleles of the genes occur, intelligence measures may be evaluated in different ways depending on assumptions, mode of the test, interpretations of the results, and the availability of repeat data from adequate samplings. In those instances where the same test has been given in successive generations, it has been found that significant differences occurred. Since there is a genetic component in the expression of intelligence, one would not expect such rapid modifications in multigenic traits to be expressed. The significant increases in human longevity, earlier maturation, and increasing average height have also occurred rapidly. These characteristics certainly have been modulated by external factors, even though each is ultimately encoded in genic DNA. A similar effect of genetic and environmental interactions clearly operates in the expression of intelligence potential. Each group of people varies in its range of IQ performance, and whether or not an average value distinguishes the group as a whole for any particular measurement, there still is the inescapable fact that *overlapping* of these ranges is the constant feature of the system. Within each group there are individuals distributed throughout a broad range of variation, which results in part from genetic and in part from environmental modulations. We all are *not* equal in potential, and cannot be, because of the high levels of genetic diversity of cross-breeding populations. The unanswered question thus remains. We are not certain about the significance of the average differences between groups that have been demonstrated, although we are certain that variation does characterize each group of people.

Prospects for Improvement

There are three general approaches that are available to regulate some portion of the genetic expressions in human populations. These are eugenic, euphenic, and genetic engineering programs.

Eugenics

The term **eugenics** first was coined by Sir Francis Galton (1822-1911), a cousin of Charles Darwin, to identify improvement of the human species through selective breeding. The term has earned some disrepute because of distortions and abuses of this concept by racists, fascists, and reactionaries. There are several current developments for eugenic programs, including *sperm storage banks for artificial insemination, cell cloning* and *nuclear transplantation* techniques, and *genetic counseling*.

Human sperm can be stored at 4°C and used subsequently in artificial insemination, a practice in increasing use today. The eugenic prospects reside in the possibility for storing sperm from particular individuals deemed to have genetically desirable traits. There are obvious difficulties once we include value judgments in the procedures, and also because we are ignorant of the genetic basis for many human behavioral qualities. There are similar moral, social, and personal difficulties in initiating programs for improvement through cell cloning or nuclear transplantation techniques. It is technically possible in cloning studies to produce replicates of an individual organism from single cells removed from the body and grown under optimal controlled conditions in culture. This has been accomplished for some vertebrate species, and could be extrapolated for human purposes. Since each of the cells from one individual has an identical set of alleles derived by mitotic divisions from the original fertilized egg, each body cell of one individual theoretically is identical in its genes. A similar situation could result by transplanting a nucleus from a body cell into an egg and then permitting the egg to develop to adulthood under the direction of the diploid chromosome constitution of the inserted nucleus. If the same individual furnishes nuclei for transplantation into a number of egg cells, then each of these could develop into an identical replica of the nuclear donor. Such methods have proven successful using frogs and toads. Both cloning and nuclear transplantation provide theoretical possibilities for mass production of replicated individuals. Whether or not this ever will happen is a question for the future and the value judgments of generations as yet unborn.

Genetic counseling (see Chapter three) provides opportunities for families to plan more accurately in situations where some element of genetic risk may be involved. Pregnancies can be aborted if tests reveal a genic or chromosomal defect in the embryo or fetus, and a couple may produce the number of children they desire without the occurrence of an afflicted child in this group. If the genetic history of the families indicates the presence of some undesirable disease or deformity, the couple may base their family planning on a more accurate knowledge of the probability of risk in producing such afflicted children.

Euphenics

The improvement of the individual by biological means constitutes programs of **euphenics**. The correction of genetic problems using preventive and therapeutic medicine can be more effective and more rapid than eugenic measures. The transplantation of organs such as kidney, heart, cornea of the eye, and others, has opened new corrective possibilities for many people with either genetic or nongenetic difficulties. The use of artificial organs and limbs also is included in the general concept of euphenics.

A classical example of euphenics in ameliorating a genetic affliction is provided in cases of *hemophilia*. Individuals who are born with a deficient system for manufacturing necessary blood-clotting factors can be helped by transfusions and by treatments with exogenously provided clotting agents. Another example concerns children who are born with *phenylketonuria*, a recessive genetic disease in which the amino acid phenylalanine is not completely metabolized because of an enzyme deficiency. The amino acid and some toxic derivatives accumulate in the bloodstream and have a deleterious effect on the central nervous system. Serious mental retardation may result, but this can be prevented by substituting a special low-protein diet for the usual milk, eggs, and other foods high in phenylalanine. Once past the critical period of early childhood, mental retardation will not develop in children who were provided with the proper diet. In both these genetic diseases there is no change in the genetic constitution of the individual, but the characteristic can be modified to subdue or mask the gene expression.

A wide variety of genetic afflictions can be reduced in severity or "cured" by appropriate treatment with enzymes, hormones, antigens, and other chemicals that cannot be manufactured or utilized by the afflicted individuals. Although the symptoms can be modified, there is no change in genes, so that the deleterious alleles can be transmitted to progeny. The people in subsequent generations also can lead useful, productive, and comfortable lives with euphenic treatment, even though they carry the genetic instructions for some affliction. Such people are fully equivalent to the large numbers of individuals who could not function in our society without the aid of eyeglasses. We do not reject anyone who requires an aid for better vision, any more than we are apprehensive about other forms of genetically based human variations. Indeed, we look forward optimistically to a continuing series of medical improvements and biological knowledge that will help to relieve symptoms in the future of some present-day incurable diseases and malformations.

Genetic Engineering

Unlike the eugenic and euphenic measures, **genetic engineering** provides a means for *directly modifying encoded genetic information*. By methods that still are primarily in the experimental stage, it has been possible to accomplish the transfer of particular genes from one cell into a new host cell. The newly incorporated DNA then proceeds to direct new expressions according to the introduced genetic instructions. It may be possible to replace a defective gene that cannot effect the manufacture of an essential enzyme with an allelic alternative that is capable of guiding the synthesis of the specific protein needed for a healthy existence. Once this can be achieved then the affliction will have been "cured," not by diet or transfusion or some euphenic modification, but directly within the individual by one's own (altered) genetic information. Furthermore, the functional new allele will be transmitted to the progeny instead of the mutant allele that it replaced.

For How Long Will We Exist?

There is, of course, no answer that can be given in absolute terms to such a question. If we examine the fossil record we find that no species survives forever; each was and is destined for extinction at some future time. But extinction does not necessarily imply that the genes of the species vanish from the Earth or that continuity is terminated. Many species have become extinct by evolving into new species, and presumably this occurred during the millions of years of hominine improvement that led ultimately to modern *Homo sapiens* from our ancestral extinct species. We also will continue to evolve, if we do not annihilate ourselves and the whole planet in the meantime, and someday *Homo sapiens* probably will become extinct. The duration of *Homo erectus* was about 500,000 years; Neanderthal types existed for about 100,000 years, and only recently became extinct as a recognizable group. We may have first appeared in our present form about 50,000 years ago, but we cannot predict for how much longer we will continue to exist.

The evolution of hominines led to entirely new dimensions of evolutionary change, unique to our species group. We can fly, we can swim, we can live in a submarine under the oceans or in a capsule out in space, and we can live in arctic cold and tropical heat; we are different from all other creatures in this respect. The diversity of human life and experience is a direct consequence of the genetically directed potential for the biological attributes of intelligence and tool-making skills. The twin directives of cultural and biological evolution have lifted our species to a totally new dimension of evolutionary possibilities. Whether or not we have a future rests in large part within ourselves, and no other species is capable of such a level of self-direction. Optimist and pessimist thus will answer the question posed above with different guesses.

Suggested Readings

Davis, K, Population. *Scientific American, 209* (September 1963), 62.

Deevey, E. S., Jr., The human population. *Scientific American, 203* (September 1963), 194.

Davis, B. D., Prospects for genetic intervention in man. *Science, 170* (1970), 1279.

Dobzhansky, Th., The present evolution of man. *Scientific American, 203* (September 1960), 206.

Dobzhansky, Th., *Mankind Evolving*. New Haven, Conn.: Yale University Press, 1962.

Dobzhansky, Th., On the evolutionary uniqueness of man. In *Evolutionary Biology*, Vol. 6 (eds., Th. Dobzhansky, M. K. Hecht, and W. C. Steere). New York: Appleton-Century-Crofts, 1972, pp. 415–430.

Garfield, S. R., The delivery of medical care. *Scientific American, 222* (April 1970), 15.

Greep, R. D., Prevalence of people. *Perspectives Biol. Med., 12* (1969), 332.

Gurdon, J. B., Transplanted nuclei and cell differentiation. *Scientific American, 219* (December 1968), 24.

Hardin, G., Nobody ever dies of overpopulation. *Science, 171* (1971), 527.

Hauser, P. M., The census of 1970. *Scientific American, 225* (July 1971), 17.

Kangas, L. W., Integrated incentives for fertility control. *Science, 169* (1970), 1278.

Langer, W. L., The black death. *Scientific American, 210* (February 1964), 114.

Langer, W. L., Checks on population growth: 1750–1850. *Scientific American, 226* (February 1972), 92.

Lederberg, J., Genetic engineering and the amelioration of genetic defects. *Bioscience,* 20 (1970), 1307.

Li, C. C., Human genetic adaptation. In *Essays in Evolution and Genetics* (eds., M. K. Hecht and W. C. Steere), New York: Appleton-Century-Crofts, 1970, pp. 545–577.

Linder, F. E., The health of the American people. *Scientific American,* 214 (June 1966), 21.

MacNeish, R. S., The origins of New World civilization. *Scientific American,* 211 (November 1964), 29.

McArthur, N., The demography of primitive populations. *Science, 167* (1970), 1097.

Murray, B. G., Jr., What the ecologists can teach the economists. *New York Times Magazine* (December 10, 1972), p. 38 ff.

Natural History Magazine, World population. January 1970, pp. 60–62.

Neel, J. V., Lessons from a "primitive" people. *Science, 170* (1970), 815.

Pimentel, D., Population regulation and genetic feedback. *Science, 159* (1968), 1432.

Simpson, D., The dimensions of world poverty. *Scientific American,* 219 (November 1968), 27.

Sjoberg, G., The origin and evolution of cities. *Scientific American,* 213 (September 1965), 54.

Taylor, C. E., Population trends in an Indian village. *Scientific American,* 223 (July 1970), 106.

Wynne-Edwards, V. C., Population control in animals. *Scientific American,* 211 (August 1964), 68.

Index

Abortion, 56, 58, 149, 160–162, 164–165, 252, 262
Abstinence, sexual, 149–151
Adaptation, 77–78
Adolescence, sexual outlet during, 207–211, 214–216
Adrenal gland, 103, 115, 248
Affectional systems, 198–201
Afterbirth. *See* Placenta
Aggression, 190
   in stickleback behavior, 177
Aging, effects of, 207–209, 214–216
Allantois, 127–128, 132
Alleles, 30–31. *See also* Genes
Alpha male and female, 188, 193
Amniocentesis, 66–67
Amnion, 66–67, 131–132, 134
Amniotic fluid, 66–67, 128, 142
Anatomy of sex organs, developmental, 117–121
   in females, 106–110
   in males, 93–99
Androgen, 93, 102–103
   in embryo, 118
   in females, 103
   in sex differentiation, 247
   in sexual disorders, 248
Animal behavior, 169–171
   development of, 197–201
   dominance in, 188–190
   patterns of, 171–185
   territoriality and, 187–190
Animal contacts, 204, 214–216
   in female sexual outlet, 211
   in male sexual outlet, 215
Animals, estrus in, 115
   length of gestation in, 130
   reproductive cycles of, 116–117
   sex determination in, 25–27, 51, 85–87

Anthropoid, 233–234
Anthropologists, studies by, 229–231, 240–245
Anus, 98, 109, 211
Artificial insemination, 163, 265
Associative learning, 184–185
Australopithecine, 234, 236–237
Autoerotic activities, 220–221
Autosomes, 8–9
  inheritance involving, 30–32, 34–37, 39–42
  structural changes in, 59, 61–62, 74–76
  types of, in humans, 49–50, 56, 58
Averages, statistical, 205–206
A-Z test, 129, 132

Baboon, dominance in the, 189, 193, 196–197
  as prehuman model, 239–244
Baby, birth of, 137–144
Barr body, 52–53
  in prenatal diagnosis, 68
  in sexual anomalies, 55–56
Basal body temperatures, 115, 162
Beach, F. A., 204, 224
Behavior, 169
  animal patterns of, 171–185
  controls in, 169–171
  cultural differences in human, 227–229, 244–246
  definition of, 169
  development of, 197–201
  human sexual, 203–221
  mating, 194–197
  social, 187–193
Birth, 137, 141–144
  defects, 53–69
  multiple, 142–143
  position of fetus before, 141–142
  premature, 136–137
  rate, 251–255, 257–264
  ratio of males and females, 26
Birth control, 149–162, 251–252
  devices, 152–154
  effectiveness of, 152
  pills, 112–113, 155–159
  rhythm method of, 150–151
Bisexuality, 218
Blastocyst, 124, 126, 134
Blood clots and the pill, 157
Breast, 103, 115, 144–145
  cancer, 221, 223
Breech birth, 142

Cancer of reproductive organs, 223–224
  and the pill, 157
Castration, 93–94, 102–103
Celibacy, 151, 260
Cell division, mitotic and meiotic, 20–22
Central nervous system, and behavior, 169–171, 174, 176–177, 180–181
  hormonal output of, 103–106, 112–113
  and lactation, 145–146
  and sexual functioning, 100–102, 116–117
Cervix, 107–109
  cancer of the, 223
  changes during labor, 141–142
Cesarian section, 123
Childbirth. *See* Birth
Childlessness, 162–164
Chimpanzees, 185, 241
Chorion, 127–128, 131–132

Chromosome, 7–23, 48–52
   aberrations, 51–63
   changes, 59–63, 74–76
   numbers, 8, 50–52
   in sex determination, 9–10, 25–27, 49–53, 86–87
   structure, 7–8
Chromosome anomalies, 51–63
   autosomal, 56–62
   prenatal tests for, 68
   sex, 51–57
   significance of, 62–63
Circumcision, 100, 224
Clitoris, 107, 109
   embryonic development of, 118–120
Coitus, 106, 123, 204, 209, 211–217
   interruptus, 151
Conception, *see* Fertilization
Condom, 152
Contraception, 149, 152–159
Contraceptives, for females, 154–159, 164
   for males, 158
   relative effectiveness of, 152
Convergence, evolutionary, 243
Copulation, *see* Intercourse, sexual
Corpus luteum, during menstrual cycle, 110–113, 115
   during pregnancy, 128–130
   see also Progesterone
Corpus spongiosum, 100
Cortex of brain, 176–177
Courtship, animal, 177–178, 181
Cowper's gland, 94, 98–99

Darwin, Charles, 77, 79
Death rate, 251, 254–255
Diaphragm, vaginal, 152, 153

Differentiation, of gametes, 23–25, 80
   of sex, 25–27
   of sex organs, 117–121, 247–248
Dilatation stage of labor, 141–142
Distribution curve, statistical, 205–206
Divergence, evolutionary, 243
DNA (deoxyribonucleic acid), 11–18
Dominance, gene, 31–36, 39–47
   and genetic risk, 63–66
Dominance hierarchies, 188–193
Douche, 152, 154–155
Down's syndrome, 56–62
   and age of mother, 59–62
   prenatal diagnosis of, 68
   trisomy-21 in, 58

Ectopic pregnancy, 126
Effectors in behavior, 170–171
Egg, 3, 9, 106–115, 125–126
   evolution of the, 80–85
   formation of the, 21–24
   see also Oocyte; Oogenesis; Ovum
Ejaculation, 99–101
Ejaculatory ducts, 94, 98–101
Embryo, development of, 129–135
   evolution of, 82–85
   implantation of, 124, 126
   sexual differences in, 117–120, 247–248
   see also Fetus
Endocrine gland, 170–171, 180–181
Endometrium (Uterine lining),

INDEX

and implantation, 111, 124, 126
in menstrual cycle, 111–114
Epididymis, 94, 96, 98–99
Erection, 99–101, 172–173
  of clitoris, 107
  of penis, 99–101
  as reflex, 100–101, 172–173
Erogenous zones, 210–211
Estradiol, see Estrogen
Estrogen, 106
  in menstrual cycle, 111–115
  in parturition, 143
  in the pill, 112, 155–157
  in pregnancy, 126–129
  see also Progesterone
Estrus, 115–117
Eugenics, 265–266
Eukaryote, 4
Eunuch, 94
Euphenics, 266–267
Evolution, 71–87
  of egg and embryo, 80–85
  of gametes, 78–80
  and natural selection, 77–79
  of separate sexes, 81
  of sex, 78–87
  of sex chromosomes, 86–87
  of sex roles, 229–246
Excitement phase, 209–212
Exhibitionism, 220

Fallopian tube, 107–108
  development of, 118–119
  and fertilization, 123–125
  surgery involving, 159–160
Family, in monkeys, 200–201
  planning for, 258
  size of, 251–252, 262–263
  as social unit, 196
Female reproductive anatomy, 107–108

anomalies, 56–57
development of, 117–121, 246–247
hormones and, 110–115
sex organs of, 107–108
Fertile period, 117, 150–151
Fertility, 115–117
  and childlessness, 162–164
  drugs to aid, 164
  and infertility, 149, 159–160, 162
Fertilization, 3, 23
  in animals, 91–92
  in humans, 123–125
Fetishism, 219
Fetus, 127
  defects in the, 59, 139–140
  development of the, 135–138
  membranes, 136
  prenatal diagnosis of, 66–68
  size at birth of, 137
  see also Embryo
Fixed-action behavior, 174
Foam, contraceptive, 152, 155
Follicle, ovarian, 108–110
  during menstrual cycle, 110–113
Follicle-stimulating hormone (FSH), in males, 102–105
  in menstrual cycle, 111–113, 146
  in pregnancy, 128–129
  and the pill, 155–158
  releasing factor, 103–106, 157
Forebrain, 176–177
Foreskin, 100
  and penile cancer, 224

Gametes, 3, 80
  evolution of, 78–80
  formation of, 21–25

Gender identity, 246—249
Genes, 6—18
  autosomal, 32, 39—42
  sex-influenced, 35—37
  sex-linked, 33—35, 42—46
  on Y chromosome, 37—38
Genetic code, 14—16
Genetic counseling, 68—69
Genetic defects, 63—69
Genetic engineering, 267
Genital ducts, 96—101, 107—108
  development of, 118—119
Genitalia, 93—95, 98—101, 107—109
  embryonic development of, 117—121, 247
  female, 107—109
  male, 93—95, 98—101
  of sexual anomalies, 53—56, 247—249
Germ cells, *see* Gametes
Gestation, 129
  length of, 129—130
Glans penis, 100
Gonadotrophins, and contraceptives, 155, 157—158
  in females, 111—114, 128—129
  in males, 101—106
  pituitary, 104
Gonad, 21, 26—27
Gonorrhea, 221—222
Gorilla, 235—237
  behavior of, 241

Haploid phase, 81—82
Harlow, H., and M., 197—202
Hemophilia, 43—44
Hermaphroditism, 27
Heterosexuality, 213, 217
  and affectional systems, 200
Hominid, 233—234, 236—237

Hominidae, 234, 236—237
Hominine, 234, 236—240
Hominoid, 233—234, 236
Homo erectus, 237—240
Homo habilis, 237
Homo sapiens, 236—237
Homophile, 217
Homosexuality, 102, 217—218
Hormones, 93, 169—171
  and behavior, 169—171
  gonadotrophic, 101—106, 111—114, 128—129
  and the pill, 112, 155—158
Human chorionic gonadotrophic (HCG), 128, 130
Human chromosomes, 49—50
  anomalies involving, 51—62
Hunting, prehuman, 239—240
  and male superiority, 229—231
Hymen, 107
Hypothalamus, 103, 145
  and behavior, 169—171, 175—177
  and lactation, 145—146
  and pituitary, 103—105, 111—113

Implantation of ovum, 123—126
  ectopic, 126
  prevention of, 156, 158
Impotence, 101
Incest, 218—219
Incompatibility, gamete, 80
Infant-mother love, 198—199
Infectious diseases, 221
Infertility, 162—163
Inheritance patterns, 29—47
  autosomal, 30—32, 39—42
  and risk, 63—68
  sex-influenced, 35—37
  sex-limited, 37

sex-linked, 31—35, 42—47
Y chromosome, 37—38
Intelligence, 185
  in human evolution, 229—240
Intercourse, sexual, 206—210, 214—217
  resolution period following, 209—210
Interstitial cells, 96, 102—104
Interstitial-cell-stimulating hormone, *see* Luteinizing hormone
Intrauterine contraceptive devices, 152—154
Inversion, chromosome, 75
IUCD, IUD, *see* Intrauterine contraceptive devices

Kinsey, A. C., studies by, 205—216
Klinefelter's syndrome, 52—55

Labia, 107—109, 120—121, 211
Labioscrotal swellings, 118—121
Labor, stages of, 141—144
Lactation, 104, 144—147, 170
Lactose, digestion of, 147—148
Langur, 241—242
Learning, 173, 183—185, 245
Lesbian, 218
Lippes loop, 154
Love, *see* Affectional systems
Luteinizing hormone (LH), in contraception, 155—158
  in lactation, 145—146
  in males, 102—105
  in menstrual cycle, 111—113
  in pregnancy, 128—129
  releasing factor, 103—105, 155—158
  site of production of, 102—103

Macaque monkeys, 241—243
  studies using, 197—201
  social systems of, 189, 193
Male reproductive anatomy, 93—101
  hormones and, 101—106
Mammals, 92—93, 232—233
  behavior of, 179—180, 188—189
  gestation period of, 129—130
  primates as, 232—233
  societies of, 192—193
Mammary gland, 92, 144—146
Margulies spiral, 154
Marriage, patterns, 251—252, 260
  and sexual outlet, 208, 211—217
Marsupial mammals, 84—85, 92, 233
Masochism, 220
Masters, W. H., and V. E. Johnson, 204, 224—225
  study of human sexual response, 209—212
Masturbation, 220
  as sexual outlet, 204, 211, 213—216
Maternal age and Down's syndrome, 56, 59—60, 68
Maternal love, 198
Mating behavior, 194—197
Mechanical contraceptive devices, 152—154
Medical genetics, 63—69
Meiosis, 21—23
Menarche, 106
Menopause, 106—107
Menstrual cycle, 106—115
Menstruation, beginning of, 106
  and birth control pills, 113—114
  cessation of, in pregnancy, 128

INDEX

duration of, 108
hormonal influences on, 111–115, 146
in nonhuman animals, 115
Milk secretion, 144
and flow, 144–146
Miscarriage, 62
Mitosis, 18–21
Money, John, studies by, 248–249
Mongolism, *see* Down's syndrome
Mortality, infant, 255–256, 260–261
Motivated behaviors, 175–183
Multiple births, 110, 142–143
Mutation, 17–19, 72–73

Natural selection, 77–79
Neurohormones, and behavior, 169–171, 176–177
in contraception, 157–158
in lactation, 144–146
in menstrual cycle, 111–114
and testicular function, 103–105
Nocturnal emissions. *See* Sex dreams
Nursing, *see* Lactation

Older people, sexual outlet of, 207–211, 214–216
Oocyte, 23–24, 106–111, 123–124
Oogenesis, 23
Operant conditioning, 184
Oral contraceptive, 155–159, 164
Orgasm, 101, 211
frequency of, 206–211
Orgasmic phase in sexual response, 209–212

Ovary, 106–113
embryonic development of, 117–119
function of, 106–113
ovarian follicles of, 108–111
ovarian phase changes in, 111–113
Oviduct, *see* Fallopian tube
Ovulation, 108–113
and conception, 123–124, 128, 150, 152, 162–163
induced, 114, 117, 123, 150
during lactation, 146
and the pill, 113, 155–158
Ovum, 23–24
development of the, 108–110
fertilization of, 123–125
lifespan of, 106–107
Oxytocin, 103, 143, 145

Pap smear test, 223
Parallel evolution, 243
Parturition, 141
Paternal love, 200–201
Pedigree analysis, 38–45
Peer love, 200
Pelvis and upright posture, 237
Penis, 94, 98–101
cancer of, 224
erection of, 99–101, 172–173
in sexual response, 100–101, 210–211
embryonic development of, 118–120
Phermone, 175
Pill, the, 154–159
dangers in use of, 157
effectiveness of, 152
and fertility, 164
hormones in, 155
for men, 158–159

Pincus, G., studies by, 155
Pituitary gland, 101–105
  activity in lactation, 104
  hormones of, 101–105
  relation to hypothalamus, 103–104, 169–170
Placenta, 127–129, 138–142
Plateau phase of sexual response, 209–212
Pongid, 233–237
Population control, 251–265
Postorgasmic response, 209–212
Potency, 93–94, 101, 159
Pregnancy, 127–141
  difficulties during, 138–141
  duration of, 131
  ectopic, 126
  hormonal activity in, 127–130
  prevention of, 149
  signs of, 132
  tests, 129, 132
  tubal, 125
Prenatal diagnosis, 66–69
  and determination of risk, 63–66
Prenatal development, 129–138
  sexual differentiation during, 117–121, 247
Primary sex characteristic, 26–27
Primates, 232–235
  societies of, 241–244
Progestagen, *see* Progesterone
Progesterone, 156
  in birth control pills, 155–159
  chemistry of, 156
  in menstrual cycle, 111–115
  in pregnancy, 126–130, 143
Progestin, 155
Prokaryote, 4
Prolactin, 103–104
  and milk flow, 144–145
  releasing factor, 145
Prostaglandin, 99, 123, 141–142, 158
Prostate gland, 94, 98–99
  cancer of, 224
Puberty, 94–95, 106–107

Races, human, differences among, 46, 65–68, 140–141
Ramapithecus, 236, 239
Reasoning, 173, 185
Receptors in behavior, 170–172
Recombination, gene, 5, 31–37, 73–74
Reflex behavior, 172–173
Refractory period, 209, 212
Reinforcement, 184
Releasing factors, 103–105
  as contraceptives, 157–158
  of hypothalamus, 103–105, 145
  synthetic, 157
Reproduction, 3–6, 194–196
  asexual, 4–5, 78
Resolution phase, 209–212
Response, human sexual, 209–212
Rh factor, during pregnancy, 139–141
Rhesus monkey, studies of, 197–201
Rhythm method, 150–152

Sadism, 220
"Safe period," 150–151
Scrotum, 94, 98–99
Secondary sex characteristic, 27
Semen, 98–99, 101
Seminal fluid, 98
Seminal vesicles, 98–99

Seminiferous tubules, 93–96, 102–103
Sex, -change operation, 218
  chromosomes, 9–10, 25–27, 49–59
  differentiation of, 25–27
  hormones, 101–105, 110–116, 155–159
  nature of, 3–6
  organs, see Ovary; Testis
  roles in society, 226–232, 239–249
Sex determination, 25–28
  chromosomes involved in, 25–26, 48–51
  in humans, 48–57
  patterns of, 25–26
  systems of, 85–87
  time of, 26
Sex dream, 213–216, 220
Sex drive, 102, 212
Sex hormones, 93, 102–106, 111–115, 155–159
Sex-influenced inheritance, 35–36
Sex-limited inheritance, 37
Sex-linked genes, 33–35
Sex-linked inheritance, 31–35, 42–46
"Sex skin," 117
Sexual behavior, human, 203–221
Sexual cycles, 111–117, 179
Sexual identity, 246–249
Sexual intercourse, see Intercourse, sexual
Sexual outlet, human, 205–216
Sexual response cycle, 209–213
Sexual variations, 217–221
  in object choice, 217–220
  in sexual aim, 220–221

Sickle cell anemia, 65–66
  genetics of, 65–66
  mutation causing, 18–19
Social behavior, 187–202
  development of, 197–201
  mating as, 194–197
Sperm, 93–97
  count in ejaculate, 101, 162
  delivery of, 96–99
  fertilization by, 123–125
  lifespan of, 95
  maturation of, 95
  production of, 24–25, 93–96, 103–105
  storage of, 98
  storage banks, 265
  X and Y containing, 9–10, 26–27
Spermatid, 24, 95–96
Spermatocyte, 24, 95–96
Spermatogenesis, 24, 95–96
Spermatogonia, 24, 95
Spermatozoa, see Sperm
Spermicide, 154
Stereotyped behaviors, 171–175
Sterility, 162
Sterilization, 91, 159–160
Steroid, 156
Stickleback, three-spined, 177–178
Synthetic hormones, 155, 157
Syphilis, 222

Tay-Sachs disease, 63–65
  inheritance of, 63–65
  occurrence of, 65
  tests for, 66–68
Testis, 93–99
  descent of, 94–95
  development of, 117–120
  hormones of, 102–103

Testosterone, 156
  effects of, 102–103
  production of, 93, 105
Tiger, L., 246, 250
Transcription, gene, 15–17
Translation, gene, 15–17
Translocation, chromosome, 75–76
  in Down's syndrome, 59, 61–62
Transsexual, 218
Transvestite, 219
Trimesters of pregnancy, 135–138
Turner's syndrome, 53, 55–57

Umbilical cord, 127, 141–142
Urethra, 98–99
  development of, in fetus, 118–121
  female, 109
  Male, 98–99
Uterus, 107–109
  in childbirth, 141–144
  and contraception, 153–154
  contractions of, in labor, 141–142
  development of, 115, 118–120
  in menstrual cycle, 110–115
  during pregnancy, 123–129

Vagina, 107–109, 124
  acidity of, 100
  in childbirth, 141–142
  and contraception, 152–155
  development of, 118–121
Vas deferens, 94, 96–99
  sterilization involving, 159–160
Vasectomy, 93, 159–160, 263
Vasopressin, 103, 145, 169–170
Venereal disease, (VD), 221–222
  condom and, 152
Voyeurism, 220
Vulva, 107–109, 118–121

Wet dreams, see Sex dreams
Withdrawal of penis, 151
Womb, see Uterus
Women, sex role of, 231–232, 240, 244–246
  sexual outlet of, 207–217

X chromosome, 9–10, 25–27, 49–59
  evolution of, 86–87
  and gender, 247–249
  inheritance, 33–35, 42–47

Y chromosome, 9–10, 25–27, 49–55
  evolution of, 86–87
  and gender, 247–249
  inheritance, 37–38
Yolk sac, 84, 128, 132–133

Zero population growth, 262–264